Longitudinal Regression Models for Population Dynamics

This book is an invaluable resource for all researchers interested in understanding practical applications of longitudinal regression methods. This book aims to equip readers with practical applications of longitudinal regression using case studies derived from administrative datasets such as the Australian government's Person Level Integrated Data Asset (PLIDA). In providing practical examples concerning longitudinal regressions, a variety of administrative data is used, including examples not only from PLIDA but also from other sources like the Institute for Employment Research, Germany. These practical examples based on these data sources are generalisable to other data sources that readers may come across and use in their analytical research.

The author emphasises the applications of longitudinal regression methods, using substantial empirical illustrations, designed to help users of social research and data analysis to better analyse and understand linked administrative datasets. This book discusses numerous SAS procedures, such as PROC GLM for estimating fixed-effects linear models, PROC LOGISTIC for estimating fixed-effects logistic regression models, PROC PHREG for estimating fixed-effects Cox regression models, and PROC GENMOD for estimating fixed-effects Poisson regression models. Readers learn about the critical need for the proper handling of longitudinal data, including issues related to data privacy, anonymisation, and ethical considerations. By providing real-world case studies and practical examples, this book bridges the gap between theoretical knowledge and practical implementation, offering valuable guidance for researchers and practitioners.

The primary audience for this book comprises social researchers and academics who engage in advanced analytical research, utilising longitudinal datasets across a variety of domains, including income, employment, health, social security, and education. This book is also well suited for statisticians, demographers, public policy analysts, and graduate students who focus on longitudinal studies to understand trends and patterns in population dynamics.

Taylan Akay, PhD, currently works at the Department of Defence as a data specialist. He also works at the University of New South Wales (UNSW) Canberra at the Australian Defence Force Academy (ADFA) as a postdoctoral senior research associate, advancing AI ethics for autonomous weapon systems. Dr Akay's career spans rigorous analytic roles across the Department of Defence, Social Services, and Home Affairs, where he has crafted scalable machine learning models, driven cloud migration strategies, and implemented enterprise data governance frameworks. Based in Canberra, he earned a PhD in economics at the Royal Melbourne Institute of Technology (RMIT) University and a master's in applied finance at Monash University.

Longitudinal Regression Models for Population Dynamics

Taylan Akay

CRC Press
Taylor & Francis Group
Boca Raton London New York

CRC Press is an imprint of the
Taylor & Francis Group, an **informa** business
A CHAPMAN & HALL BOOK

First edition published 2026
by CRC Press
2385 NW Executive Center Drive, Suite 320, Boca Raton FL 33431

and by CRC Press
4 Park Square, Milton Park, Abingdon, Oxon, OX14 4RN

CRC Press is an imprint of Taylor & Francis Group, LLC

ISBN: 978-1-032-98665-4 (hbk)
ISBN: 978-1-032-98661-6 (pbk)
ISBN: 978-1-003-59986-9 (ebk)

DOI: 10.1201/9781003599869

Typeset in Palatino
by codeMantra

Access the Support Material: www.routledge.com/9781032986616

Contents

1

Foundations of Longitudinal Data Analysis: Designs, Trajectories, and Causal Leverage

The term cross-sectional data describes observations collected simultaneously from multiple subjects. This data type provides a snapshot of the relationships among population traits at that particular time. For example, cross-sectional survey results could show the typical income, level of education, and employment conditions of subjects in a specific demographic group for a particular year. Cross-sectional studies find common application in the social sciences, economics, and medicine, as they are typically easier to conduct and economically cheaper than their longitudinal counterparts. Nevertheless, the absence of a temporal aspect renders cross-sectional data incapable of illustrating shifts or patterns over time.

Contrary to cross-sectional data, longitudinal data features several measurements of the same subjects at different intervals. This approach enables researchers to monitor changes and trends within subjects or groups, capturing temporal dynamics often hidden in cross-sectional studies. Therefore, the defining characteristic of longitudinal data is its capacity to observe how subjects' traits develop, providing insights into the trajectories of the population. This is in contrast to time-series data that generally consists of extended observations for a limited group, concentrating on population-level temporal dynamics instead of individual, subject-specific risk factors. Longitudinal cohort data entails multiple measurements on the same subjects over a shorter timeframe, enabling researchers to track changes within subjects and pinpoint effects related to age, period, and cohort.

Longitudinal data play a crucial role in population research and various other fields, as they enable the modelling of trajectories, forecasting of future outcomes, and managing unobserved heterogeneity among subjects, in other words, differences in characteristics among subjects within a population group. By examining the progression of subjects over time, longitudinal analysis helps in disentangling the genuine impacts of population characteristics, yielding a deeper understanding of causal relationships and trends.

A typical longitudinal dataset notation follows a matrix and an equation definition. Let $Y_{i,t}$ denote the value of a response variable for the i-th subject at a time t. A typical longitudinal dataset, $Y_{i,t}$, is measured for multiple subjects $i = 1, 2, ..., N$ across multiple time points $t = 1, 2, ..., T$:

DOI: 10.1201/9781003599869-1

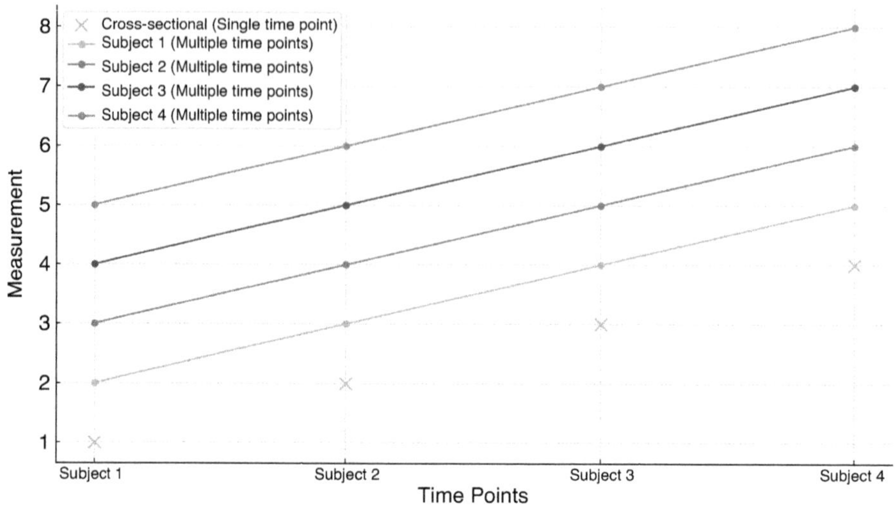

FIGURE 1.1
Cross-sectional vs. longitudinal data.

$$Y = \begin{bmatrix} Y_{1,1} & Y_{1,2} & \cdots & Y_{1,T} \\ Y_{2,1} & Y_{2,2} & \cdots & Y_{2,T} \\ \cdot & \cdot & \cdots & \cdot \\ \cdot & \cdot & \cdots & \cdot \\ \cdot & \cdot & \cdots & \cdot \\ \cdot & \cdot & \cdots & \cdot \\ Y_{N,1} & Y_{N,2} & \cdots & Y_{N,T} \end{bmatrix}$$

A row in this matrix represents the repeated measurements for a single subject, while each column corresponds to observations at a specific time point.

In longitudinal data, we observe a series of repeated measurements for each subject over time. For example, let $Y_{i,t}$ represent the value of the outcome variable at time t. The model for the individual subject i's outcome over time can then be expressed as follows:

$$Y_{i,t} = \beta_0 + \beta_1 t + b_i + \varepsilon_{i,t}$$

where

β_0 is the intercept, representing the baseline outcome value for a subject,
β_1 captures the effect of time on the outcome, shared across all subjects,

b_i represents the random effect for subject i, capturing how this subject's baseline value differs from the overall population (individual-specific deviations), and

$\varepsilon_{i,t}$ is the error term at time t, capturing random noise in the measurement for the subject i.

The term $\beta_1 t$ shows the effect of time on the outcome variable, allowing us to capture the subject-level trends. By tracking the same subject over time, we can observe the unique trajectory of $Y_{i,t}$ as it evolves with time. For instance, a healthcare investigation could illustrate a subject's blood pressure trends. The model allows us to monitor how each subject's blood pressure fluctuates, which may be affected by factors such as age, medical treatment, or lifestyle changes. This enables population researchers to assess whether the blood pressure result rises, falls, or remains stable for each subject.

Identifying group-level trends in longitudinal data involves analysing repeated measurements to uncover group patterns. This process is important in various fields, including psychology, epidemiology, and genetics, in understanding temporal dynamics and making informed estimates. Group-level trends emerge by aggregating subject-level changes across the population. In longitudinal data analysis, the average trajectory of a group can be estimated, for example, using the fixed effects β_0 and β_1, which represents the group-level trend in the outcome variable over time, $E(Y_{i,t}) = \beta_0 + \beta_1 t$. This equation gives us the expected outcome value at time t, which applies to the population. By estimating β_1, we can determine whether the group-level trend is positive (e.g., outcomes improve over time), negative (e.g., outcomes deteriorate over time), or stable.

Modelling the temporal dimension of longitudinal data helps population researchers in grasping causal connections by examining how variations in independent variables affect the outcome over time, such as a covariate recorded at one time point relating to an outcome evaluated at a later time point, utilising the temporal sequence to imply causality. Understanding causality in longitudinal data entails pinpointing causal links from information gathered over extended periods, a task that is inherently difficult due to the observational characteristics of these studies. Further, tracking changes within the same subject also allows population researchers to account for time-invariant confounding variables, thereby enhancing their understanding of causal inferences. A standard model for causality in longitudinal data analysis is the fixed-effects (FE) model, where the time-varying covariate $X_{i,t}$ represents an intervention that changes over time, $Y_{i,t} = \beta_0 + \beta_1 X_{i,t} + b_i + \varepsilon_{i,t}$. Here, β_1 captures the effect of the covariate $X_{i,t}$ on the outcome $Y_{i,t}$, controlling for subject-specific differences (through b_i) that do not vary over time. For example, if $X_{i,t}$ represents whether a subject received treatment, the coefficient β_1 measures the causal effect of the treatment on the outcome over time.

Longitudinal data also enables the prediction of future outcomes by modelling the trajectory of the outcome variable over time. This task is crucial in population research, where understanding temporal patterns can inform decision-making and policy development. The complexity of longitudinal data, characterised by high dimensionality, irregular measurement intervals, and potential missing data, necessitates sophisticated approaches, such as random-effects (RE) models. If we assume a linear relationship between time and the outcome variable, we can then use the estimated parameters β_0 and β_1 to predict the outcome at future time points, such that $\widehat{Y}_{i,t+h} = \widehat{\beta}_0 + \widehat{\beta}_1(1+h)$, where $h$represents the number of future time points ahead for which we want to make a prediction. This is particularly useful in estimating trends, such as predicting future health outcomes, financial metrics, or social behaviours based on the observed trajectories presented in the data. For example, in a potential study on income mobility, if we know that $\beta_1 > 0$, we can then predict an upward trajectory for a subject's income in future years. While this approach offers a robust framework for predicting future outcomes, challenges remain, such as managing missing data, ensuring model interpretability, and adapting models to different contexts.

Longitudinal and cross-sectional data are two essential forms of data utilised in population research. Each data form serves specific objectives, yet the time-related nature of longitudinal data makes it particularly effective for examining changes and trends. This aspect is vital in population research, as it helps in identifying trends at both subject- and group-level population, understanding causal relationships, and estimating future outcomes relevant for social policymakers across government agencies, non-governmental organisations, and research institutions.

Using longitudinal data analysis has become a fundamental aspect of evidence-based decision-making in governmental policy, social research, healthcare, and economics. Through observing the shifts within subjects or groups over extended periods, longitudinal data provides a profound and nuanced perspective on behaviours, outcomes, and trends that cross-sectional data cannot yield. For example, in health service management, this method of scrutiny enables the observation of disease development, the evaluation of the success of long-term interventions, and the prediction of patient outcomes. In economics, it reveals trends in income mobility, employment shifts, and the impact of policy actions on individuals over time.

To equip readers with a practical understanding of real-life examples of longitudinal data, the following discussion provides a high-level overview of cases where longitudinal data significantly influence evidence-based decisions, drawing on examples from Australia's Person Level Integrated Data Asset (PLIDA) and the Household, Income and Labour Dynamics in Australia (HILDA) survey.

One of the most significant uses of longitudinal data is to improve public policy and guide programmatic efforts. For instance, Australia's PLIDA is a vital resource for the Australian Government, providing insights from a wide range of administrative datasets, including health, education, employment, and censuses. The PLIDA dataset enables policymakers to observe the lasting impacts of policy interventions on specific population groups and adjust programmes based on these findings. For instance, longitudinal data derived from the PLIDA dataset is utilised to analyse the lasting impacts of government social assistance programmes on Australians' employment paths. The amalgamation of diverse datasets, including those related to taxation, community support, and medical records, fosters a well-rounded view of a subject's life experiences, critical for unravelling the connections between various socio-economic factors and their resultant impacts. By tracking changes in employment status over time, policymakers also assess the effectiveness of specific programmes, such as JobKeeper, and make informed decisions about the potential continuation, modification, or termination of such initiatives based on data-driven evidence.

Mathematically, the impact of a policy intervention on employment status $Y_{i,t}$ can be modelled as

$$Y_{i,t} = \beta_0 + \beta_1 \text{ Policy}_t + \beta_2 X_{i,t} + b_i + \varepsilon_{i,t}$$

where

$Y_{i,t}$ represents the employment status of subject i at time t,

Policy_t is a binary variable indicating the presence of the policy intervention at time t,

$X_{i,t}$ represents subject-specific covariates (e.g., educational attainment, age, or previous income),

b_i captures the random effect specific to subject i, and

$\varepsilon_{i,t}$ is the error term.

Such a model allows policymakers to isolate the effect of the policy intervention while controlling for individual-specific factors and external influences over time.

In healthcare, the study of longitudinal data is essential for tracking patient outcomes over time, evaluating the effectiveness of treatment methods, and identifying risk indicators associated with multiple health issues. The HILDA Survey provides researchers and healthcare policymakers with longitudinal insights into the health and well-being of Australians, monitoring changes in health-related behaviours, access to medical services, and health outcomes among different demographic segments. For example, an application of longitudinal data within the healthcare sector is the examination of chronic disease trajectories. By persistently tracking the health conditions of subjects across lengthy spans, researchers can recognise trends that may signal disease advancement, as well as the success or lack thereof of treatment methods.

Furthermore, extended studies have the potential to enhance the evaluation of healthcare measures by comparing health outcomes before and after such measures are applied, thus allowing for inferences about causal influences. For instance, researchers aim to elucidate the long-term impact of a novel medical treatment on patient health. They could use a longitudinal regression model to analyse the relationship between the treatment $X_{i,t}$ and the health outcome $Y_{i,t}$ as follows:

$$Y_{i,t} = \beta_0 + \beta_1 X_{i,t} + b_i + \varepsilon_{i,t}$$

where

$Y_{i,t}$ represents the health outcome of patient i at time t,

$X_{i,t}$ indicates whether the patient received the treatment,

b_i is the random effect capturing the variability between subjects, and

$\varepsilon_{i,t}$ is the error term.

By analysing changes in the outcome $Y_{i,t}$ over time, researchers assess the effectiveness of the treatment while accounting for subject differences.

Longitudinal data studies, such as the HILDA Survey, are similarly essential for comprehending the dynamics of the labour market, tracking household income, employment, and labour trends over extended time frames. They reveal information regarding job transitions, salary hikes, job reliability, and the repercussions of economic instability on employment conditions. By consistently monitoring the same subjects over time, economists investigate the fundamental causes and effects of employment variations, including job loss, career transitions, and the impact of technological advancements on specific industry sectors. For instance, economists leverage longitudinal data from the HILDA Survey to analyse the enduring consequences of unemployment on earnings pathways by examining subjects' wage trends before and after periods of unemployment. Longitudinal data enable the identification of wage scarring, a phenomenon defined by prolonged unemployment that has adverse effects on future earnings. Such analytical methods are crucial for developing labour market policies to mitigate the adverse effects of unemployment.

The following equation can be used to analyse how unemployment affects wages as

$$\text{Wage}_{i,t} = \beta_0 + \beta_1 \text{ Unemployed}_t + \beta_2 X_{i,t} + b_i + \varepsilon_{i,t}$$

where

$\text{Wage}_{i,t}$ represents the wage of subject i at time t,

Unemployed_t is a binary variable indicating whether the subject is unemployed at time t,

$X_{i,t}$ represents other time-varying covariates, such as education or industry of employment,

b_i captures the random effect specific to subject i, and

$\varepsilon_{i,t}$ is the error term.

The HILDA Survey is an essential resource for tracking income mobility across Australia, enabling population researchers to delve into the mechanisms by which individuals rise or fall within the income hierarchy and the factors influencing these shifts, such as educational background, family influences, and labour market conditions. By analysing income trajectories, economists can uncover upward or downward mobility trends and assess the impact of government initiatives aimed at reducing poverty and addressing inequality, which in effect would equip policymakers to design interventions that tackle chronic poverty instead of fleeting economic issues.

The hierarchical structure of longitudinal data features repeated measures within subjects, who can be further categorised into larger groups such as classrooms or communities. This nested arrangement adds another layer to the traditional multi-level data framework, positioning longitudinal data as a unique and powerful tool across a range of fields. This multi-level framework also enhances data analysis at different levels of aggregation and can accommodate complex relationships among variables.

Analysing longitudinal data typically involves assessing the first- and second-order properties of the data and its correlation structures over time. These structures are crucial for understanding growth patterns and changes at both group and subject levels. This dynamic nature requires specialised modelling techniques to address multiple modalities and scales. The temporal scale may vary depending on the specific characteristics of the phenomena. For example, intervals may be evaluated in days or weeks during clinical trials to identify shifts. In contrast, in social or behavioural studies, they might be assessed in months or years. Another important consideration of longitudinal studies is the number of temporal points incorporated. While using two or three time points provides limited insights, involving four or more time points allows for modelling more complex, non-linear trajectories. A greater number of time points improves the model's capacity to represent the response variable's actual trajectory accurately. However, increasing the time points also raises the complexity and financial costs of the longitudinal data collection methods.

In a typical multivariate data setup, each subject is depicted by a distinct row in a dataset with various measurements recorded across the columns. This structure is commonly known as the wide format. For example, suppose a research investigation collects each participant's four consecutive measurements of a variable like blood pressure. In that case, the resulting data matrix can then be illustrated as shown in Table 1.1.

In this table, several assessments (e.g., two separate blood pressure readings) are aligned horizontally across columns that correspond to each subject. The main advantage of this layout is that it allows for a visual assessment of

TABLE 1.1

Sample Multivariate Longitudinal Data Structure

ID	Measurement 1	Measurement 2	Measurement 3	Measurement 4	Age	Gender
1	120	125	118	130	40	Male
2	110	112	115	113	35	Female
…	…	…	…	…	…	…
…	…	…	…	…	…	…
…	…	…	…	…	…	…

TABLE 1.2

Sample Univariate Longitudinal Data Structure

ID	Time	Measurement	Age	Gender
1	1	120	40	Male
1	2	125	40	Male
1	3	118	40	Male
1	4	130	40	Male
2	1	110	35	Female
2	2	112	35	Female
…	…	…	…	…
…	…	…	…	…
…	…	…	…	…

each subject's development across different measurement periods. However, there are certain limitations, as temporal factors are represented indirectly through the arrangement of the columns, and this format becomes cumbersome when dealing with more complex models that include time-dependent variables, like blood pressure or weight.

The univariate data structure, commonly called the long format, on the other hand, is employed in longitudinal studies due to its ability to provide flexibility in treating time as an explicit variable. In this structure, each participant is depicted through multiple records, with each record relating to a specific moment in time. For example, the same dataset seen in the multivariate scenario in a univariate format is shown in Table 1.2, which shows each temporal measurement point is listed as an individual row, with the relevant data for each participant at every temporal point recorded within a single column. This approach enhances the effectiveness of modelling temporal effects, especially when temporal points are unevenly distributed or where covariates show temporal fluctuations.

Another critical aspect of longitudinal studies is the classification of data as either balanced or unbalanced. In a balanced design, every subject design is evaluated at the identical set of time intervals, with no instances of missing data. For example, if all subjects in a study are assessed at the same four time points (e.g., months 1, 2, 3, and 4), the design is considered balanced. On the other hand, the data is often unbalanced, meaning that not all subjects have measurements at each time point due to attrition, missed appointments, or delayed participation in the study. For example, if some subjects are assessed only at months 1 and 3, while others are evaluated at months 1, 2, 3, and 4, the data is considered unbalanced. Unbalanced datasets add complexities to the analytical process; however, modern statistical techniques, including longitudinal regression models, are specifically tailored to tackle this challenge.

Longitudinal data take various forms, contingent upon the characteristics of the measurements and the statistical requirements, for example, diverse classifications of longitudinal data are fundamentally based on the structural composition and objectives of the research endeavour, and comprehending these classifications is crucial for selecting the most suitable analytical methodologies. The following discussion outlines the primary types of longitudinal data, encompassing event history, repeated measures, and time-series data, while explaining their mathematical foundations and practical applications. Event history data is about the temporal dynamics and occurrences of specific events across time. This data type is especially advantageous in population research that investigates the duration preceding the manifestation of a particular event, such as marriage, job cessation, or disease onset. Event history data analysis is helpful in studying time-to-event data and assessing how covariates (e.g., age, education, treatment) influence the likelihood of an event occurring over time. Event history data is also subjected to analysis employing survival analysis methodologies.

Repeated measures data involves multiple measurements of the same outcome variable over time for each subject. It is typically used to track changes within subjects in response to interventions or time. Repeated measures data is usually analysed using longitudinal regression models, which allow for both FE (common to all subjects) and RE (unique to each subject). These models are ideal for managing the within-subject correlation that arises from repeated observations of the same subjects. Repeated measures data is the focus of this book. While time-series data refers to repeated observations over time for a single subject or entity, in a longitudinal context, multivariate time series arise when multiple subjects or entities are observed over time. Each subject's time series can be analysed separately, or multivariate techniques can be employed to study interdependencies.

Mathematically, time-series data is represented as

$$Y_{i,t} = f\left(Y_{i,t-1}, Y_{i,t-2}, \ldots, Y_{i,t-p}\right) + \varepsilon_{i,t}$$

where

$Y_{i,t}$ is the outcome variable for a subject i at time t,

$f(\cdot)$ is a function of the past values of the outcome variable (up to lag p), and

$\varepsilon_{i,t}$ is the error term.

Autoregressive (AR) models and autoregressive integrated moving average (ARIMA) models are prevalent methodologies for examining time-series data within longitudinal research. They effectively address the interdependence observed among successive time intervals and facilitate forecasting forthcoming values by utilising historical data. Growth curve models, on the other hand, represent a specific category of longitudinal analysis that emphasises modelling an outcome's trajectory over a temporal continuum. These models prove to be especially advantageous in examining developmental alterations or the progression of various conditions, such as the growth patterns of children or the cognitive deterioration observed in geriatric populations.

The simplest form of a growth curve model is the linear growth model, where the outcome variable $Y_{i,t}$ is modelled as a function of time as

$$y_{i,t} = \beta_0 + \beta_1 t + u_i + \varepsilon_{i,t}$$

where

β_0 is the intercept, representing the baseline level of the outcome at time 0,

β_1 is the slope, representing the rate of change of the outcome over time,

u_i is the random effect for subject i, capturing subject variability in growth trajectories, and

$\varepsilon_{i,t}$ is the error term.

Longitudinal data involves repeated measures from the same subjects. This results in correlated observations and if not adequately accounted for, results in biased estimates and invalid inferences. While intra-subject correlation poses challenges in longitudinal data analysis, it also offers opportunities to explore complex behavioural hypotheses by blending inter-subject differences with intra-individual dynamics. For instance, RE models offer significant advantages in addressing intra-subject correlations in longitudinal data, making them a preferred choice for population researchers. These models allow for the specification of alternative covariance structures, which can address both intra- and inter-subject heterogeneity, providing flexibility in modelling the dependencies prevalent within the data.

Although longitudinal data provides profound insights into temporal changes, it introduces obstacles for population researchers, such as managing missing data, ensuring data quality, and selecting the appropriate statistical models. The common challenges linked with longitudinal data can be categorised into heterogeneity, variability among subjects, errors in data restructuring, irregularly spaced measurements, and missing data. A prevalent misconception in longitudinal modelling is the assumption that the relationships between predictors and outcomes remain unchanged throughout the entire observation period. Nevertheless, these relationships often exhibit temporal variations; neglecting to identify or accurately model these fluctuations can obscure significant temporal dynamics and lead to erroneous interpretations. Another aspect population researchers need to be aware of is that transitioning from wide to long data formats requires meticulous attention. Mistakes in data reshaping, such as misaligned timestamps or incorrectly matched variables, can severely compromise the accuracy of the analysis. Moreover, in real-world research situations, subjects frequently exhibit different measurement intervals, resulting in data points that are not evenly distributed. Conventional statistical approaches often face challenges with these unbalanced designs, making it essential to employ techniques, such as RE modelling, which can adeptly handle these inconsistencies. Finally, yet importantly, missing data represents a prevalent challenge within longitudinal research studies, and the methodologies employed to address such missing data can profoundly influence the analytical validity of the findings. Three principal mechanisms characterise missing data:

1. Missing Completely at Random (MCAR): The absence of data transpires in a manner that bears no relation to the observed or unobserved data. For instance, a subject may fail to attend a scheduled appointment due to an arbitrary conflict in their calendar.
2. Missing at Random (MAR): The likelihood of data being absent is contingent upon the observed data rather than the missing data. For example, subjects of advanced age may exhibit a higher propensity to neglect a subsequent survey. However, the justifications for their absence are not intrinsically linked to their health outcomes.
3. Missing Not at Random (MNAR): The absence of data is influenced by the unobserved data. For instance, subjects experiencing deteriorating health may withdraw from a research study, indicating that their missing data is associated with their unobserved health status. Analyses of longitudinal data frequently operate under the assumption of MAR, thereby permitting specific modelling strategies to manage missing data while minimising bias effectively.

Cross-sectional datasets are commonly analysed through regression models that fail to consider temporal correlations. In a cross-sectional analyses, one might employ a straightforward linear regression model, such as $Y_i = \alpha_0 + \beta_1 X_i + \varepsilon_i$, where Y_i is the outcome variable for subject i, X is a predictor, and ε_i is the error term. Such model assumes that subjects' observations are independent of each other. The defining feature of a longitudinal data model, on the other hand, is its ability to examine changes over time within subjects and variations over time among groups. For example, longitudinal models can calculate regression coefficients at the subject and population levels, specifically for subjects.

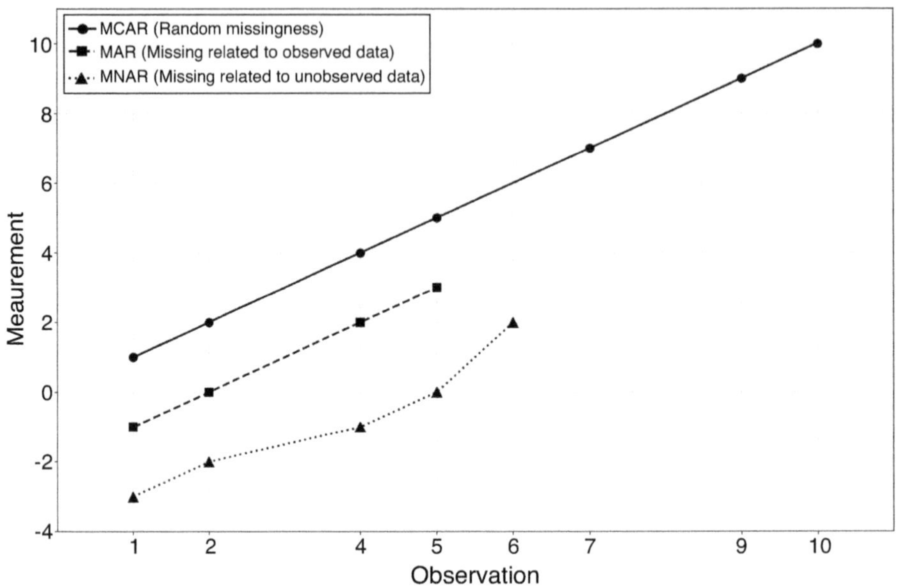

FIGURE 1.2
Types of missing data: MCAR, MAR, and MNAR.

Longitudinal data analysis offers a refined and dynamic insight into changing processes, hence functioning as a reliable framework for examining transformations, causation, and variability at individual and collective levels. Longitudinal data requires the application of specialised statistical approaches owing to the intrinsic correlation found in measurements from identical subjects, the increased correlation when measurements are taken closely in time as opposed to those taken over lengthier durations, and the tendency for variances to evolve in longitudinal data over time. These potential correlation and variation patterns can collectively create a complex covariance structure. It is crucial to consider this covariance structure to obtain valid statistical conclusions. As a result, traditional regression and ANOVA models may produce misleading outcomes, as two essential parametric assumptions might not hold, for example, independent observations and homogeneity of variances, might not hold. As mentioned, examining longitudinal cohort data requires the application of specialised statistical approaches to address the inherent correlations linked with repeated measures. Failing to recognise this correlation could lead to underestimating model variance and diminished standard errors, which may result in incorrect interpretations.

The primary strategies for handling correlated errors encompass FE, RE, and generalised estimating equations (GEE) models. Each methodological option possesses unique benefits, rendering it appropriate for diverse research designs and aims. These methodologies underscore the importance of longitudinal regression modelling in elucidating temporal dynamics, whether the focus is on population-level averages or trends specific to individual subjects. The judicious selection of the appropriate model empowers researchers to harness the comprehensive nature of longitudinal data, thereby facilitating a nuanced comprehension of the progression of key variables over time.

Understanding the assumptions about subject effects and their relationship with regressors in longitudinal models is crucial for accurately estimating and inferring these models. These assumptions significantly influence the choice of estimation techniques and the credibility of the conclusions drawn from analyses of longitudinal data. For instance, FE models assume that all subjects in the study maintain a consistent relationship between variables, focusing on estimating the effects of predictors that remain uniform across subjects or groups, effectively control for confounding factors by addressing within-subject elements that do not change over time, as they remove the influence of time-invariant characteristics through intra-subject comparisons. On the other hand, RE models incorporate subject-level differences by introducing subject-specific random intercepts and slopes, thus permitting variations in baseline levels and rates of change among subjects. These models adeptly capture individual heterogeneity and provide insights at the population level while simultaneously modelling the correlation present within repeated measures. Longitudinal regression modelling methodologies are essential for analysing data that manifests variability across temporal dimensions and diverse entities. The principal differentiations among fixed effects, random effects, and GEEs are based on their respective assumptions concerning individual effects and their interactions with regressors. Each methodological framework offers advantages and limitations that substantially influence its applicability, depending on the structural characteristics of the data and the research aims (Table 1.3).

Determining which model to adopt should reflect the specific attributes of the dataset and the nature of the research question. FE models are considered optimal when controlling for unobserved heterogeneity correlated with the independent variables, which is essential. Conversely, RE models are preferred when such correlations are assumed to be non-existent, thus allowing for incorporating a broader spectrum of variables. GEEs are useful in estimating population-averaged effects, rendering them advantageous for analysing correlated data without necessitating subject-specific conclusions. Different methodologies carry their field of relevance, and selecting the best model calls for a profound grasp of the foundational assumptions and compromises present in every research method.

TABLE 1.3

Key Differences Across the FE, RE, ME, and GEE Methods

	Fixed Effects (FE)	Random Effects (RE)	Generalised Estimating Equations (GEE)
Assumption	Subject effects are posited to be correlated with the regressors	Individual effects are presumed to be uncorrelated with the regressors	GEE models estimate population-averaged effects rather than subject-specific effects
Use case	FE models are particularly suited for analysing variations occurring within a subject over temporal intervals. These models control for constant characteristics, allowing researchers to isolate the effects of variables that exhibit temporal variability	RE models prove advantageous when the research inquiry encompasses both within- and between-subject variations	GEE is especially beneficial for analysing correlated datasets, such as those from repeated measures studies
Strengths	FE models yield consistent estimates under the condition that subject effects are correlated with the regressors	RE models permit the incorporation of time-invariant variables as predictors and can be enhanced with random coefficients to capture intricate relationships	A significant advantage of employing GEE is its robustness against misspecification of the correlation structure
Limitations	A notable limitation of FE models is their incapacity to estimate the effects of variables that remain constant over time	A potential drawback is that if individual effects correlate with the regressors, the RE model may produce inconsistent estimates, leading to biased inferences	GEE models may exhibit lower efficiency than mixed models, mainly when subject-level effects are of interest

Reflection Questions and Answers

Question 1 – How does the temporal nature of longitudinal data enhance causal inference compared to cross-sectional data?

Answer 1 – Longitudinal data provides temporal ordering of events, enabling more accurate causal inference. For example, evaluating the effectiveness of a job-training programme over several years allows policymakers to attribute employment improvements specifically to the training.

Question 2 – Why is it critical to distinguish between fixed and random effects when interpreting results from longitudinal data analysis?
Answer 2 – Fixed effects capture overall population trends, while random effects highlight individual differences. Understanding individual variability in response to educational reforms helps tailor interventions for distinct demographic groups.

Question 3 – What are the practical implications of incorrectly assuming data missingness as MCAR in a longitudinal study?
Answer 3 – Assuming MCAR is incorrect may bias policy evaluations. For example, if disadvantaged individuals drop out of a financial literacy study disproportionately, policy decisions might incorrectly assume the programme benefits all groups equally.

Question 4 – Explain how temporal variability in predictor effects might impact policy recommendations derived from longitudinal studies.
Answer 4 – Policy effectiveness can fluctuate over time. For instance, an anti-poverty intervention might initially reduce poverty but become less effective without ongoing support, necessitating policy adjustments.

Question 5 – What strategies can researchers use to validate the proper data restructuring from a wide to long format?
Answer 5 – Detailed documentation and validation scripts prevent restructuring errors. For example, employment data must be meticulously verified after restructuring to track individuals' employment statuses over time accurately.

Question 6 – Discuss how intra-subject correlation affects the precision of estimates in longitudinal analyses.
Answer 6 – Ignoring intra-subject correlation can inflate significance estimates. For instance, neglecting repeated measures of household consumption can lead to a false suggestion of substantial policy impacts.

Question 7 – Why are balanced designs often impractical in longitudinal studies, and how do modern methods address this issue?
Answer 7 – Dropouts make balanced designs rare. For instance, random-effects models effectively handle dropout in evaluating the long-term welfare policy impacts.

Question 8 – Reflect on the ethical considerations involved in managing longitudinal data, particularly regarding participant confidentiality.
Answer 8 – Ethical practices safeguard sensitive data, which is crucial for maintaining trust in long-term socio-economic studies.

Question 9 – How can incorrect interpretations of temporal dynamics in longitudinal studies influence clinical practice?
Answer 9 – Misinterpreted temporal dynamics may yield ineffective clinical guidelines. For example, misjudging the timing of vaccine effectiveness could lead to suboptimal immunisation schedules.

Question 10 – What challenges do population researchers face in predicting future outcomes using longitudinal data, and how might they overcome these?
Answer 10 – Complex covariance structures can obscure predictions. Robust models and diagnostics, such as those used to predict pension outcomes, enhance accuracy.

Question 11 – Explain why understanding covariance structures is crucial in longitudinal data analysis.
Answer 11 – Complex covariance structures can obscure predictions. Robust models and diagnostics, such as those used to predict pension outcomes, enhance accuracy. Proper covariance models inform accurate policy impacts. For instance, correctly modelling covariance is vital to understanding the dependence of repeated income measurements.

2

Administrative Longitudinal Data: Architecture, Governance, and Harmonisation

Longitudinal studies have increasingly employed survey and administrative datasets, each with unique pros and cons pertinent to the research inquiry at hand. Historically preferred for social science research, survey data has benefited from systematic data collection designed to address research inquiries, ensuring elevated data quality and pertinence. This methodological control over data collection protocols promotes uniform measurements across survey iterations, thereby mitigating measurement errors and affirming the integrity of the data. Survey data also facilitates the incorporation of participant consent, streamlining ethical considerations and confidentiality protocols.

Nevertheless, longitudinal surveys encounter challenges regarding participant retention over extended periods due to attrition, costs, and constraints within sampling frameworks, particularly in engaging eligible population cohorts or ensuring long-term participation. In contrast, administrative datasets comprise data collected during the implementation of government programmes, services with legal mandates. Unlike data derived from longitudinal surveys, these datasets are not initially gathered for research purposes but can be repurposed for academic and policy-oriented research. These datasets, typically compiled by governmental agencies, provide extensive and continuous coverage, making them amenable to comprehensive population analyses.

Administrative datasets encompass records about populations and can document event histories with a degree of temporal precision, chronicling interactions as they transpire rather than depending on retrospective accounts provided by respondents. Despite these advantages, administrative data presents several challenges. They remain vulnerable to changes in administrative policies, which may affect the consistency of the data collected. Further, there is limited control over the specific data variables collected, as these are selected to support administrative objectives rather than particular research goals.

In this chapter, we will examine the structural, ethical, and privacy dimensions of administrative data collection and its use in population research. We will concentrate on three noteworthy examples: Australia's Person Level

DOI: 10.1201/9781003599869-2

Integrated Data Asset (PLIDA), Canada's Longitudinal Administrative Databank (LAD), and Germany's Linked Employer-Employee Data from the Institute for Employment Research (IAB), referred to as LIAB.

Given its extensive and accessible documentation, particular emphasis will be placed on Australia's PLIDA.

PLIDA is an ideal illustration of the capacity for government-led data initiatives to amalgamate various administrative data sources, yielding a comprehensive longitudinal dataset, which assists population researchers in understanding Australian population dynamics. Instituted by the Australian Bureau of Statistics (ABS) in conjunction with multiple governmental entities, PLIDA synthesises health, education, income, taxation, employment, and demographic data. Its architecture is conducive to longitudinal analysis, empowering researchers to monitor transformations over time and among population groups.

The legal foundation for PLIDA's data integration is enshrined in legislative acts, specifically the Census and Statistics Act of 1905, accentuating the ABS's jurisdiction to collect and safeguard data. At the nucleus of PLIDA resides the person linkage spine, a structural apparatus that functions as a unique, anonymised identifier, thereby connecting subject records across heterogeneous datasets. This foundational framework secures data linkage while upholding data integrity and participant confidentiality. The segregation of the person linkage spine from the data repository ensures the protection of identifying information, thereby fortifying PLIDA's comprehensive privacy protocol.

The ABS implements a Fellegi–Sunter probabilistic framework, supplemented by clerical review of high-risk pairs in person linkage spine development. In its process, each administrative file (e.g., Medicare consumer directory, personal income tax) is compared based on subjects' name, date of birth, sex, and address. Then, record pairs receive a composite weight and are linked whenever that weight exceeds a threshold chosen to balance false and missed-match rates. ABS's annual quality audits publish two headline indicators: (i) precision, the share of declared links that are true and (ii) recall, the share of all true links that are found. For the 2016 Spine update, they reported an overall precision of 98.7 per cent and recall of 97.9 per cent, implying a false-match rate of about 1 in 75 links and a missed-match rate of about 1 in 50 possible links.

Integrating datasets within PLIDA promotes a multifaceted perspective on population dynamics that underpins the ability of population researchers to conduct extensive research. For example, the database interlinks information from the Australian Census, health surveys, employment records, taxation data, and other sources facilitates longitudinal analyses encompassing demographic attributes, income distribution, employment trajectories, and health outcomes of Australians. Similarly, PLIDA integrates health data with income and employment records, allowing population researchers to examine the impact of socio-economic status on health outcomes and service accessibility. Furthermore, by interlinking educational records with employment data, PLIDA enhances academic inquiries into the relationship between educational attainment and success in the labour market.

PLIDA's structural design enables the longitudinal tracking of individuals and households, crucial for examining life-cycle changes and enduring trends in socio-economic mobility. For instance, PLIDA integrates data collected for operational rather than research purposes; therefore, its underlying definitions evolve, personal income-tax thresholds are re-indexed nearly every budget cycle, Centrelink payment categories are revised when programmes merge or cease, and health service items migrate to new Medical Benefits Scheme (MBS) item numbers after clinical guideline updates. Hence, if population researchers pool waves of data without adjustment, apparent trends may reflect definitional change. Best practice is to map each variable onto a harmonised series that is stable across years, using the concordance tables the ABS publishes, for example, (i) create lookup tables in their code base, (ii) recode historical observations onto the latest scheme, and (iii) document any residual break points in metadata so that downstream users can apply sensitivity checks. This harmonisation step ensures that observed shifts, such as the rise in bulk-billed GP visits, movement between income deciles, are attributable to behavioural or policy changes rather than shifting administrative classifications.

The grey boxes in Figure 2.1 show agency-specific data feeds (Health, Social Security, Tax, and Education). All identifiers are routed through the person linkage spine, which produces a de-duplicated master key; only then are research variables assembled into the anonymised research file and released

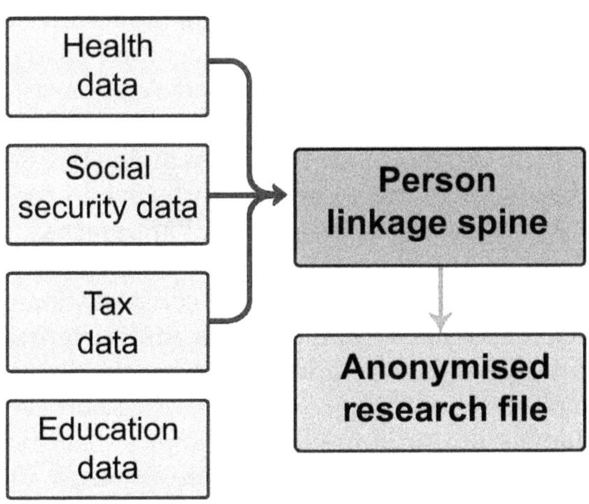

FIGURE 2.1
Integrated administrative plumbing.

to population researchers under the Five Safes Framework (safe projects, safe people, safe data, safe settings, and safe outputs).

Although PLIDA's structure exemplifies data integration, Canada's LAD and Germany's LIAB present alternative structural methodologies. LAD, administered by Statistics Canada, consolidates annual income-tax records in a longitudinal format, allowing population researchers to investigate income mobility and economic patterns over extended periods. This configuration supports life-course analysis and policy research on economic disparities and social welfare. Conversely, Germany's LIAB, overseen by the IAB, interrelates employer data with employee records to enhance the understanding of labour-market dynamics. This forked approach yields unique insights into employment practices, job stability, and wage growth. The structure of LIAB places a premium on detailed analysis of job transitions and employment conditions, thereby supporting macroeconomic research and the formulation of workforce policies.

In Australia, administrative datasets operate under an implied-consent model where data is collected initially for service delivery or compliance and, subject to legislation like the Data Availability and Transparency Act 2022 and ABS's Five Safes Framework, may be repurposed for research without re-contacting each individual. The ethical premise is that reuse delivers public value while robust de-identification, secure DataLab access, and project-by-project approval protect individuals' privacy expectations. By contrast, population surveys (e.g., the National Health Survey, NHS) rely on an opt-in consent model in which respondents are explicitly informed about research aims, data linkages, and storage arrangements, and participation is voluntary at every wave. This approach maximises participant autonomy and transparency but can introduce non-response rates and higher costs.

For longitudinal studies, the choice of consent model affects both data richness and ethical risks. Implied consent yields comprehensive, near-universal coverage yet demands strong governance and safeguards, whereas opt-in consent affords granular self-reported variables and clear participant agency but often suffers from attrition and limited sample size. Researchers must weigh these trade-offs, document consent pathways in their research protocols, and, where feasible, incorporate hybrid approaches, such as seeking explicit consent for sensitive linkages or relying on implied consent for less intrusive administrative enrichments. These considerations underscore the significance of legal and ethical frameworks in administering administrative datasets. They ensure that data is handled with respect to individual privacy while promoting public benefit.

PLIDA is fundamentally supported by principles that underscore the significance of responsible data utilisation, community trust, and ethical deliberations. As the integrator and steward of PLIDA, the ABS adheres to rigorous protocols that prioritise transparency and serve the public interest. The maintenance of ethical standards is facilitated through systematic consultations with stakeholders, which encompass data custodians, governmental entities, and the wider community. These engagements ensure that data is used to respect the rights and privacy of individuals, particularly those belonging to vulnerable demographics.

Key ethical considerations for PLIDA revolve around maintaining public trust while maximising societal benefit. First, the ABS commits to complete transparency, ensuring citizens understand what data PLIDA contains, how it is used, and the policy implications that flow from those uses. This openness is reinforced by robust ethical oversight: a dedicated governance board comprising representatives from all contributing agencies monitors PLIDA's operation and guarantees alignment with national data-ethics standards. At the same time, stringent secrecy provisions under the Census and Statistics Act 1905 protect the dataset; no information may be released in a form that could re-identify individuals, ensuring strict compliance with privacy legislation. Consistent with these controls, the ABS mandates the ethical use of data, especially where vulnerable populations are concerned, requiring adherence to all relevant laws, regulations, and professional standards. Finally, PLIDA is integrated and analysed solely for statistical and research purposes where a clear public benefit exists, so its insights inform evidence-based policy and decision-making.

In Canada, LAD operates under the Privacy Act, which outlines protocols for using personal information. The ethical employment of LAD is predicated on informed consent and the anonymisation of data to prevent re-identification. Statistics Canada ensures that the application of LAD aligns with ethical standards, fostering trust among Canadian citizens while enabling researchers to access invaluable longitudinal data for policy analysis.

LIAB is governed by the German Federal Data Protection Act, which prescribes stringent data handling and access requirements. The ethical framework in LIAB emphasises the necessity of secure data environments and anonymisation techniques, thereby balancing the demand for comprehensive, detailed data with the essential obligation to safeguard individual identities.

Large-scale linkage projects such as PLIDA occupy a delicate space between public-interest analytics and the fundamental right to informational self-determination. Australian and international guidance now converges on four framing principles: legitimacy, proportionality, transparency, and accountability. Legitimacy requires that linkage serve an articulated, socially valuable purpose rather than mere data opportunism. Proportionality means collecting only those variables essential for the research questions and retaining identifiable information for the shortest feasible time. Transparency obliges custodians to publish plain-language descriptions of how personal records flow through the linkage spine, what legal gateways are relied upon, and what safeguards buttress each step. Finally, accountability anchors the chain of responsibility: linkage units, accrediting statisticians and end-user researchers each sign enforceable undertakings that align with the Australian Privacy Principles, the GDPR's data-minimisation rule, and the ABS Five Safes Framework.

A second layer of ethics concerns equity and representativeness. Administrative systems often under-record marginalised groups, such as migrants in informal employment and itinerant populations. Population researchers have a professional duty to detect and disclose such blind spots. Finally, ethics review boards increasingly evaluate algorithmic fairness when approving linkage studies. Investigators must demonstrate that downstream models (e.g., risk scores or eligibility rules) will not disproportionately disadvantage protected classes. For illustration, consider a readmission-risk model built from linked hospital, mortality, and census data in which Aboriginal and Torres Strait Islander peoples constitute the key protected class. Prior to any analysis, the investigators may preregister that the model will be deemed inequitable if the absolute difference in false-positive rates (FPR) between Indigenous and non-Indigenous patients.

When overseeing administrative datasets, it is essential to implement anonymisation and data-protection measures. The ABS embeds multiple safeguards in PLIDA to ensure rigorous privacy protection and data integrity. First, ABS staff members involved in creating, using, or managing PLIDA sign legally binding undertakings of confidentiality, reinforcing personal accountability for guarding the dataset. Second, identifying variables are segregated from content variables and housed on separate servers so that no individual can access identity information side-by-side with detailed content data. The checklist in Table 2.1 allows readers to reference the legal basis for each step of PLIDA's linkage and analysis pipeline, ensuring that research proposals align with Australian privacy standards.

Oversight goes beyond baseline compliance, relying on the Five Safes Framework to vet researchers, control access environments, and check results before release. At the data level, the person linkage spine separates personal identifiers from analytical variables, keeping only a minimal set of keys for matching and ensuring the analytical dataset remains de-identified. Together, these measures form a comprehensive governance architecture that minimises re-identification risk while preserving the analytical power of PLIDA. Statistics Canada administers LAD via secure Research Data Centres (RDCs). The RDCs implement de-identification and stringent data access protocols to safeguard personal information. They furnish a controlled research environment in which only authorised users can access data, achieving a balance between research utility and privacy. Germany's LIAB utilises secure data rooms and remote access solutions, mandating that researchers undergo a stringent approval process. Combined with anonymisation techniques, these measures uphold the privacy standards mandated by German legislation.

While PLIDA, LAD, and LIAB exhibit exemplary privacy practices, they encounter challenges common to all extensive administrative data initiatives. For example, ensuring that data is adequately anonymised while preserving the analytical depth requisite for meaningful research can prove complex, and advancements in data science, including machine learning and AI,

TABLE 2.1

Privacy-by-Design Checklist for PLIDA

Safeguards in PLIDA	Corresponding GDPR Articles (Art)	Corresponding Australian Privacy Principles
Purpose specification register	Art 5(1)(b) purpose limitation	APP 1.2. Open and transparent management
Minimal field extraction-only variables on an approved data concordance are taken	Art 5(1)(c) data minimisation	APP 3. Collection of solicited personal information
Two-pass encryption-raw identifiers encrypted at source and re-encrypted	Art 32 integrity and confidentiality	APP 11.1. Security of personal information
Separation of functions	Art 25(1) privacy by design	APP 1.2. APP 6.1 Use or disclosure only for authorised purpose
Bloom-filter PPRL for cross-jurisdiction matches	Recital 29, Art 89(1) technical safeguards for research	OAIC De-identification guidelines (cross-referenced to APP 11)
Five Safes secure lab-air-gapped virtual desktop, no outbound clipboard, vetted output only	Art 32, Art 35 data-protection impact assessment	APP 11. Privacy Safeguard 5, secure handling of data sharing
Potential dynamic-consent portal	Art 7(3) right to withdraw consent, Art 21 right to object	APP 3.5. Option of anonymity and pseudonymity
Audit-trail logging for all table reads and outputs	Art 30 records of processing activities	APP 11.2. Breach response and accountability
Annual re-identification risk test	Art 25(2) data protection by default	OAIC Stat. Guideline 3 (de-identification)
Automatic purge of linkage keys	Art 5(1)(e) storage limitation	APP 11.2. Destroy or de-identify

introduce potential re-identification risks. These challenges necessitate that authorities oversee administrative datasets to refine their privacy frameworks, effectively and continuously mitigating these risks.

PLIDA, LAD, and LIAB serve as paradigms illustrating the transformative capacity of administrative data in population research. PLIDA's extensive integration across diverse domains, LAD's longitudinal monitoring of income and demographic variables, and LIAB's bifocal insights into the intricacies of employer-employee relations are among the administrative data resources that are indispensable in shaping policy, promoting evidence-driven research, and enhancing societal outcomes. Despite the inherent challenges of reconciling privacy considerations with the exigencies of research, these administrative data resources are indispensable in shaping policy, promoting evidence-based research, and improving societal outcomes. These longitudinal datasets have already yielded substantial policy benefits across various jurisdictions. In Australia, PLIDA facilitated the Productivity Commission's

2023–2024 Snapshot of Inequality, which examined post-COVID disposable-income trends for every household quintile; its conclusion that single JobSeeker recipients experienced the most significant real-income declines informed Treasury's May 2025 decision to increase the base JobSeeker rate and introduce an energy supplement for low-income renters. In Canada, Finance Canada's review of the LAD disclosed that upward mobility from the lowest income quintile had decreased between amongst the population, evidence that prompted the 2016 substitution of Canada's child tax benefit with the generous, means-tested benefit schemes, later recognised for driving child-poverty rates. In Germany, the linked-employer-employee LIAB panel indicated that the 2015 statutory minimum wage elevated earnings for low-wage workers, particularly women in the eastern Länder, without causing significant job losses; the Minimum Wage Commission referenced these findings in 2018 when advocating for (and achieving) an increase in the wage floor. Collectively, these instances illustrate how meticulously connected administrative data can provide prompt, evidence-based insights that influence income-support, family-benefit, and labour-market policy.

The following discussion elucidates how PLIDA enhances socio-economic research and policy formulation, scrutinising its applications across various domains and its ramifications for governmental and academic endeavours. PLIDA is employed in various socio-economic research initiatives that interrogate fundamental inquiries regarding inequality, poverty, and social mobility. In the Australian context, it is foundational to Australia's complex socio-economic research framework. Integrating data from diverse administrative repositories, PLIDA enables population researchers, policymakers, and academicians to gain insights into the complex relationships between health, education, income, and employment. PLIDA's integrated, longitudinal view of Australians' lives has yielded transformative insights across multiple policy domains. In healthcare, analyses derived from PLIDA guided the allocation of medical resources toward communities with the greatest needs, directly shaping initiatives for vulnerable populations. Social-welfare agencies likewise draw on PLIDA's evidence base to track the long-term effects of benefits and refine programmes for better outcomes. Labour-market researchers use the data to map employment trajectories, informing job-training schemes and policies that address unemployment and workforce transitions.

During the COVID-19 crisis, PLIDA's timely linkage of income, employment, and demographic variables allowed officials to gauge the real-time socio-economic impact of lockdowns and craft targeted stimulus packages. The dataset has also supported climate-health resilience: by overlaying demographic and health records with meteorological data, the ABS produced a national heat-health vulnerability map that enables authorities to protect those most at risk during heatwaves. Finally, specialised PLIDA extracts focusing on Indigenous Australians provide a nuanced picture of disparities in health, education, and economic participation, underpinning policies aimed at equity and inclusion. These applications demonstrate

PLIDA's unique ability to convert comprehensive, privacy-protected data into evidence-driven policy that benefits the nation's most vulnerable groups.

In addition, PLIDA has made substantial contributions to academic research. However, its impact is comparatively less pronounced than that of more established datasets such as the HILDA Survey and the Longitudinal Study of Australian Children (LSAC) Survey. The following outlines key academic inquiries that have utilised PLIDA datasets, the analytical methodologies applied, and the research outcomes that underscored their significance within the Australian academic landscape.

Research supported by PLIDA has yielded significant findings that have informed various policy debates and scholarly discussions. For example, investigations conducted by Welsh et al. (2021) and Butler et al. (2022) emphasised the crucial connections between socio-economic determinants and health outcomes, while analyses by Ladbrook and Hendrickson (2020) and Zajac et al. (2024) have illuminated the challenges faced by disadvantaged students and the imperative for targeted intervention strategies. The insights provided by Mowle (2020) and Biddle et al. (2019) have offered nuanced perspectives on income distribution, influencing the trajectory of socio-economic policy formulation. Biddle et al. (2019) conducted an external validation of the basic longitudinal extract from PLIDA, concentrating on the distribution and dynamics of income. While accentuating the dataset's utility, they expressed caution regarding the issues of incomplete linkages and the limitations inherent in the available income data. Butler et al. (2022), on the other hand, correlated the NHS data from 2014–2015 with the MBS/Pharmaceutical Benefits Scheme (PBS) and death registrations from PLIDA and evaluated adherence to cardiovascular treatment protocols. Their logistic regression analysis showed deficiencies in therapy utilisation and identified potential areas for improvement within healthcare policy frameworks. Cowie and Hendrickson (2020) integrated national mental health service utilisation and pharmaceutical data with educational data to evaluate how mental health conditions influenced university graduation rates. Using propensity-score matching and binomial logit regression, they discovered that severe mental health disorders impeded student completion rates. Furthermore, Welsh et al. (2021) employed data from the Census 2011, personal income tax, and MBS/PBS datasets to evaluate the progressivity of out-of-pocket healthcare expenditures. They conducted cross-sectional analyses to express out-of-pocket expenses as a fraction of income, utilising the Kakwani Index to assess progressivity. The research yielded significant insights pertinent to the discourse on healthcare equity within Australia. In another exemplary research, Ladbrook and Hendrickson (2020) examined the correlation between income support and university graduation rates among Indigenous students by integrating data from the Higher Education Information Management System with PLIDA. By employing random forest and binomial logit regression methodologies, the investigators underscored the significance of socio-economic determinants in educational achievement.

Mowle (2020) used data from the Australian Taxation Office and the Department of Social Services, which were integrated with the 2016 Census, to conduct a comparative analysis of population-level income metrics. Their findings showed consistency in income distribution across these datasets, affirming the robustness of the PLIDA framework for socio-economic research. Roder et al. (2023) investigated integrated data from the New South Wales Cancer Registries with sociodemographic and health information sourced from PLIDA with a two-stage analytical approach employing cross-sectional multiple logistic regressions, cross-tabulations, and multi-variable analyses to evaluate the features of population cohorts diagnosed with liver cancer. Their findings revealed associations between a variety of sociodemographic factors and cancer diagnoses, thereby contributing to the ongoing discourse regarding early cancer detection. Saxby et al. (2023) examined the impact of fluctuating working hours on utilising mental healthcare services. The authors implemented linear fixed-effects regression analyses using data from PLIDA's Census, Medicare records, and social security payments data. Their findings indicated that working 1–14, 15–29, and ≥30 hours weekly correlated with decreased utilisation of mental healthcare services, illuminating critical implications for labour policy and mental health strategies. Welsh et al. (2021) used data from the 2016 Census and death registrations to analyse mortality rates across varying educational attainment levels through negative binomial regression modelling. Their investigation revealed that lower levels of education were correlated with elevated mortality rates, thereby highlighting the disparities in health outcomes associated with educational inequalities. Zajac et al. (2024) utilised the PLIDA dataset to examine the impact of mental health treatment on retention rates in higher education through the application of growth-curve models that revealed mental health adversities are significantly correlated with increased higher education dropout rates, underscoring the necessity for improved mental health support within educational environments.

PLIDA-supported investigations encompassed a myriad of statistical methodologies, thereby emphasising its adaptability for a wide range of research applications. Researchers drawing on PLIDA employ a spectrum of analytical techniques that range from traditional statistical models to advanced machine-learning methods. Classical regression, including logistic, Cox, and linear fixed-effects models, remains a mainstay for estimating associations between exposures and outcomes. Complementing these approaches, Biddle and Marasinghe (2021) demonstrated the value of machine-learning algorithms such as random forests and gradient boosting to create synthetic income measures and uncover complex, non-linear patterns in the data. Cowie and Hendrickson (2020) applied propensity-score matching to reduce selection bias, thereby balancing treatment and control groups before longitudinal analysis. Finally, Zajac et al. (2024) displayed the utility of growth-curve models for tracing individual income and unemployment trajectories over time, highlighting PLIDA's capacity for dynamic modelling of life-course processes. Together, these methodologies illustrate the versatility of PLIDA as a foundation for causal inference and predictive analytics across diverse policy questions.

Robust data integration has become a strategic imperative in an era where evidence-based policymaking depends on weaving together ever-more diverse administrative records. As the landscape of administrative data evolves, obstacles related to data integration, including aligning heterogeneous data formats and ensuring consistency across various datasets, persist as significant issues.

The prospects for conducting comparative studies using PLIDA, LAD, and LIAB present stimulating opportunities for global inquiries into common challenges, including economic resilience and population health. Efforts aimed at fostering data-sharing agreements and standardised methodologies have the potential to enhance cross-national research and inform policy development. Linking administrative data across different authorities to construct a comprehensive longitudinal dataset creates privacy and consent issues, given that subjects have not expressly consented to utilise their data for research purposes. As the data environment expands, legal and ethical frameworks must evolve to align with emerging realities. In the case of PLIDA, sustaining trust within the Australian populace necessitates ongoing engagement and the continual refinement of ethical protocols. Similarly, LAD and LIAB face corresponding pressures to adapt their frameworks in response to dynamic data privacy standards and ethical usage requirements.

Blending the depth of survey instruments with the scale and timeliness of administrative records holds enormous promise for longitudinal research. Still, it demands careful design and rigorous methods. While both survey and administrative datasets provide valuable insights for longitudinal analysis, their integration requires meticulous planning and methodological rigour to balance each dataset's strengths and limitations effectively. Survey data provide depth and specificity, albeit with restricted temporal granularity and elevated costs. Conversely, administrative data contribute breadth and temporal immediacy; yet, their intrinsic purpose and consistency may limit their value. Collectively, they represent a promising pathway for comprehensive longitudinal research, yielding nuanced insights into population dynamics. A prominent strategy within this domain is using an administrative spine, a foundational population dataset derived from administrative records, which then serves as a basis for linking supplementary surveys or administrative datasets. This spine supports longitudinal studies by providing a stable and identifiable sample that can be monitored over time. However, constructing an adequate spine is inherently complex, especially in settings where unique personal identifiers are not available. Such a framework also requires rigorous quality assessments, as inconsistent record linkage may compromise longitudinal analyses. Furthermore, linked datasets raise heightened ethical and confidentiality concerns, necessitating rigorous access controls and security measures to protect individual privacy.

Yet, even the most judiciously engineered spine cannot guarantee that every population segment is captured equally. Once the survey and administrative records have been linked, analysts must confront residual coverage gaps and linkage errors, issues that, if ignored, can bias longitudinal estimates. Administrative collections seldom encompass the entire resident population with uniform probability: recent arrivals, informal sector employees, or individuals lacking permanent addresses may be inadequately recorded, and linkage inaccuracies can further diminish the coverage of specific sub-groups. Consequently, the ABS publishes calibrated analysis weights alongside its longitudinal micro-files. Each weight comprises four elements, that is, design weight, under-coverage adjustment, missed-link adjustment, and benchmarking against independent estimated resident population totals, ensuring that weighted estimates accurately reflect age-by-sex-by-state totals and Indigenous status counts, even when raw linkage fails to represent those groups adequately. Analysts should (i) utilise the provided weights in all descriptive analyses, (ii) recalibrate or post-stratify when concentrating on sub-populations that are not completely benchmarked (e.g., recent humanitarian entrants), and (iii) employ replicate-weight or bootstrap methods to account for additional variance introduced by calibration. In the absence of official weights, practitioners can mimic the ABS methodology: assess under-coverage by contrasting linked counts with independent administrative or survey controls, adjust for known linkage omissions, and benchmark to external population totals before advancing with longitudinal modelling.

Robust weighting and calibration techniques help compensate for residual coverage gaps once administrative linkages are in place. Still, sustained comparability increasingly hinges on shared, standards-based metadata that lets researchers describe, discover, and securely combine linked files across borders. International efforts are aligning towards common, machine-readable frameworks that render the interconnection of administrative microdata across different jurisdictions both technically viable and legally verifiable. The DDI-Cross Domain Integration (DDI-CDI) model expands upon the established DDI metadata family, encompassing extensive, long, multi-dimensional, and key-value data, while clearly outlining linkable identifiers, provenance steps, and transformation rules; this facilitates the documentation of a PLIDA extract, for instance, in the same format as an EU SILC panel or a Canadian LAD file, significantly easing cross-national comparative research. In addition, the ISO-certified Statistical Data and Metadata eXchange (SDMX) standard offers standardised code lists and a web-service API for the exchange of aggregated outputs and their structural metadata; organisations such as ABS, Statistics Canada, and Eurostat already distribute SDMX dataflows, allowing analysts to programmatically retrieve harmonised time-series and associate concept definitions with each variable key. Collectively, DDI-CDI for detailed micro-level descriptions and SDMX for macro-level data exchange establish a comprehensive interoperability framework that reinforces standards-based metadata as the cornerstone for secure, collaborative longitudinal studies.

Reflection Questions and Answers

Question 1 – Explain how the Fellegi–Sunter linkage framework balances precision and recall in PLIDA's person linkage spine. Why is this balance critical for longitudinal policy analysis?
Answer 1 – Fellegi–Sunter assigns probabilistic weights to record pairs on shared identifiers (name, DOB, sex, address). PLIDA sets a linkage threshold that yields≈approximately 98.7 per cent precision (few false links) and approximately 97.9 per cent recall (few missed links). High precision prevents spurious person-level trajectories that would bias, say, JobSeeker exit rates; high recall ensures actual episodes (e.g., welfare spells) are not systematically lost, which would understate programme utilisation. The chosen threshold, therefore, minimises both type I and type II linkage errors, preserving the validity of longitudinal inferences that underpin social-transfer reforms.

Question 2 – Temporal drift can distort trend analyses in administrative data. Outline a three-step harmonisation workflow and illustrate it with PLIDA's re-indexed income-tax thresholds.
Answer 2 – A three-step workflow could be:

Step 1: Build concordance tables that map every historical tax bracket to the current bracket set.

Step 2: Recode each past record's income-tax status to the harmonised scheme (e.g., Bracket 3–2006–Bracket 2–2024).

Step 3: Flag residual break points in metadata for sensitivity checks. In PLIDA, applying this workflow prevents a nominal shift of taxpayers into lower brackets (caused by indexation) from being misread as downward mobility; ensuring observed income-distribution changes reflect economic reality, not tax-code drift.

Question 3 – Contrast "implied consent" and "opt-in consent" in the context of PLIDA versus a national health survey. What are the trade-offs for representativeness and respondent autonomy?
Answer 3 – PLIDA's implied-consent model repurposes service data under the Data Availability and Transparency Act 2022 without re-contacting individuals; representativeness is near-universal but hinges on robust de-identification and governance. By contrast, opt-in surveys must secure explicit participation, granting high autonomy yet suffering attrition and non-response bias (e.g., under-representation of renters or migrants). Thus, implied consent maximises coverage and longitudinal reach, whereas opt-in consent maximises informed choice but risks skewed population weights.

Question 4 – This chapter cites PLIDA data driving an increase in Australia's JobSeeker base rate. Describe the causal logic linking administrative evidence

to policy change and outline one methodological safeguard that strengthens this logic.
Answer 4 – PLIDA tracked disposable-income trajectories by quintile post-COVID, revealing disproportionate losses among single JobSeeker recipients. Policymakers inferred that base-rate adequacy, not labour-supply incentives, explained hardship. A methodological safeguard is difference-in-differences: comparing JobSeeker recipients to a stable control group (e.g., aged pensioners) before and after supplement expiry isolates the policy effect from macro shocks, bolstering the claim that PLIDA evidence warranted the benefit rise.

Question 5 – Discuss how DDI-CDI and SDMX create an "end-to-end interoperability stack". Give a concrete use case for cross-national labour-market research.
Answer 5 – DDI-CDI provides machine-readable micro-level metadata (identifier structure, provenance, variable mapping) for datasets such as PLIDA or LIAB. SDMX offers code lists and APIs for aggregated outputs (e.g., quarterly unemployment). A researcher can document LIAB micro-variables in DDI-CDI, harmonise them to PLIDA equivalents, and then publish the SDMX time series of job-separation rates for OECD comparison. Consumers programmatically join SDMX aggregates while retaining DDI-CDI links back to microdata for replication, seamlessly traversing micro-to-macro layers.

Question 6 – Why are calibrated analysis weights essential when using administrative datasets such as PLIDA for estimating Indigenous unemployment? Outline the four components of the ABS weighting scheme.
Answer 6 – Administrative under-coverage (e.g., mobility, remote addresses) and missed links deflate counts of Indigenous Australians. Calibrated weights correct this by:

- Design weight (inverse inclusion probability),
- Under-coverage adjustment (aligns to external controls, e.g., Census counts),
- Missed-link adjustment (inflates records to offset linkage failures), and
- Benchmarking to estimated resident population totals. Applying these weights yields unbiased unemployment estimates critical for Closing-the-Gap policy monitoring.

Question 7 – How did LAD evidence influence the Canada Child Benefit (CCB) redesign? Identify two longitudinal indicators from LAD that policymakers likely monitored post-reform.
Answer 7 – LAD showed falling upward mobility for bottom-quintile families, prompting conversion of the Canada Child Tax Benefit into the means-tested CCB in 2016. Post-reform, policymakers monitor child-poverty rates (share below LIC-AT) and income-mobility transitions (movement from

bottom to middle quintile within 5 years). Improvement in both indicators validates CCB's targeting and informs indexation decisions.

Question 8 – LIAB studies found no significant employment loss after Germany's 2015 minimum-wage introduction. Describe one identification strategy that could support this finding and discuss a potential limitation.
Answer 8 – A matched difference-in-differences pairs low-wage workers (treatment) with similar higher-wage workers (control) across pre- and post-2015 periods, estimating employment probability changes. A limitation is that if treated firms adjust hours or hire subcontractors, employment effects may be hidden due to composition bias. To address this, researchers should test robustness using firm-level synthetic controls/outcomes based on hours worked.

Question 9 – Outline the Five Safes Framework applied to PLIDA and illustrate how "Safe Outputs" is enforced when releasing heat-health vulnerability maps.
Answer 9 – Safe Projects (public-value justification), Safe People (accredited researchers), Safe Data (de-identified, minimised variables), Safe Settings (secure ABS DataLab), and Safe Outputs. For heat-health maps, Safe Outputs requires aggregation to the SA2 level, suppression of cells with <20 residents, and disclosure review of geospatial overlays before release, preventing re-identification while still guiding emergency services targeting.

Question 10 – Design a concise Five Safes access plan for linking State birth-registration data to PLIDA. Specify one safeguard under each of the Five Safes dimensions that would satisfy the State Registrar of Births, Deaths and Marriages (RBDM) and the ABS DataLab.
Answer 10 –

Five Safes Pillars	**Concrete Safeguard in This Linkage Project**
Safe projects	The project scope is limited to estimating low-birth-weight cohorts' long-run health and education outcomes; a written statement of expected public benefit has been lodged with RBDM and ABS.
Safe people	All team members hold current ABS microdata accreditation, have completed RBDM privacy induction, and sign individual deed-of-confidentiality addenda for vital-statistics data.
Safe settings	Linkage and analysis were undertaken in ABS DataLab; no unit-record data was exported, and two-factor authentication and an air-gapped virtual desktop were used.
Safe data	Birth records stripped of name strings and complete addresses before departure from RBDM; linkage relies on hashed Medicare number, DOB, and gender.
Safe outputs	Cell suppression and perturbation rules were applied; no table was released with fewer than five births in any cross-classification; re-identification risk was assessed with the ABS Confidentiality Tool before clearance.

3

Synthetic Panels for Methods: From Data Engineering to Estimation Strategy

By the conclusion of this chapter, readers will be equipped to articulate, in straightforward terms, how each of the three synthetic case-study files was constructed to reflect its real-world administrative equivalent while adhering fully to the Five Safes privacy framework introduced in Chapter 2. This chapter guides readers in creating essential derived variables, most notably the binary Wage Gap Indicator and interaction terms like Sector x Promotion that will appear again later in Chapter 5. This chapter provides a glimpse into the analytical journey ahead: Case Study 1 contributes to the fixed-effects (FE) linear and conditional-logit models of Chapter 5, Case Study 2 informs the random-effects (RE) of Chapter 6, and Case Study 3 supports the generalised estimating equations (GEE) of Chapter 7. Throughout this process, readers will discover why specific predictors disappear under an FE design yet reappear in RE frameworks, effectively connecting the practical data tasks of this chapter to the methodological trade-offs discussed in Chapter 1.

To enable readers to engage directly with realistic-looking micro-records while honouring the privacy-by-design principles outlined in Chapter 2, each case-study file is a fully synthetic dataset where no row corresponds to an actual person or business, yet attention was given in generating the synthetic data so the joint distributions of age, income, education, and other key variables replicate those observed in the original administrative sources.

These case studies are designed to equip readers with the essential skills for conducting thorough empirical research. They will help readers understand and interpret insights related to policy while also enhancing their understanding of how regression outputs can reveal significant patterns and trends in population research. By actively engaging with these case studies, readers will develop the ability to navigate complex longitudinal datasets, apply appropriate statistical methods, and critically assess the implications of their findings for informing public policy decisions related to socio-economic issues.

Case Study 1 – Understanding Wage Inequality

Wage inequality pertains to the disproportionate income distribution across diverse demographic cohorts arising from gender, education, occupation, or

DOI: 10.1201/9781003599869-3

regional inequities. Wage inequality has the potential to perpetuate more extensive social and economic disparities, thereby constraining opportunities and reinforcing cycles of disadvantage.

Case Study 1 endeavours to elucidate the methodology for implementing an FE model utilising SAS Studio to examine social inequality in wage distributions. It will scrutinise how much variables such as age, education, and occupation contribute to wage disparities over a temporal continuum. By employing a synthetic dataset modelled after the Linked Employer-Employee Data (LEED), the German Federal Employment Agency provided, readers will acquire hands-on experience in constructing, estimating, and interpreting FE models. FE estimation is ideal for this panel dataset because many key confounders, such as innate ability, motivation, or regional opportunity structures, are unobserved but stable over time. By using each subject as its control, the FE model eliminates bias from these time-invariant characteristics, isolating the causal impact of time-varying predictors like age, sector, or promotion.

In this case study, the synthetic dataset emulates the LIAB dataset, which includes 5,000 employees tracked over 5 years, representing linked data from employers and employees.

Because the data contain no true identifiers, they can be distributed outside secure DataLab environments without triggering the safe settings, restrictions, or the implied-consent provisions discussed in Chapter 2. This approach preserves statistical utility for teaching and code development, allows for the public sharing of replication scripts, and eliminates re-identification risk, demonstrating how synthetic generation complements the governance safeguards that apply to the PLIDA, LAD, and LIAB extracts used in this book.

This case study is crafted to serve as an empirical connection that links the extensive data preparation detailed in Chapter 4 with the FE estimation methodologies that will be thoroughly examined in Chapter 5. Following the process of reshaping the synthetic wage longitudinal data into a long format, we also take the crucial step of constructing a key derived variable, specifically, the binary Wage Gap Indicator, which is designed to indicate whether a worker's annual wage is situated below the median wage for their specific gender-by-occupation-by-education category.

In Chapter 5, we will estimate two FE models using this dataset: a linear FE model where the hourly wage is the continuous dependent variable and a conditional FE logit model where the binary *Wage_Gap_Indicator* denotes whether a subject falls below their demographic-peer median wage in a given year. Together, these models capture both mean-level changes and distributional risk, allowing for a multifaceted analysis of wage inequality. Preliminary estimates suggest that *Age* carries a statistically significant within-subject wage premium, while *Sector, Promotion,* and *Region* contribute little once subject differences are taken into account. This insight anticipates a broader empirical argument developed in Chapter 5, where education emerges as the most robust safeguard against wage disadvantage.

TABLE 3.1

Case Study 1 Dataset Structure

Variable Name	Description
ID	Unique identifier for each subject
Year	The year of observation
Wage	Annual gross wage in EURO (continuous)
Gender	Binary indicator for gender (0 for Male and 1 for Female)
Age	Years (continuous)
Job_Tenure	Years with the current employer (continuous)
General_Education	Levels: "Presently Attending Schools" "Secondary School Certificate" "Certificate of Ten-Grade School" "Intermediate School Certificate" "University Entrance Qualification" "No Secondary School Certificate"
Vocational_Qualification	Levels: "Apprenticeship" "Trade and Technical School" "Bachelor's Degree" "Master's Degree" "Doctor's Degree" "No Vocational Qualification"
Occupation	Levels: "Managers" "Professionals" "Technicians and Associate Professionals" "Clerical Support Workers" "Service and Sales Workers" "Skilled Agricultural, Forestry, and Fishery Workers" "Craft Related Trades Workers" "Plant and Machine Operators, and Assemblers" "Elementary Occupations" "Armed Forces Occupations"
Firm Size	Levels: "Small" "Medium" "Large"
Sector	Levels: "Industry" "Services" "Agriculture"
Region	Levels: "Baden-Württemberg" "Bayern" "Berlin" "Brandenburg" "Bremen" "Hamburg" "Hessen"

(*Continued*)

TABLE 3.1 (*Continued*)

Case Study 1 Dataset Structure

Variable Name	Description
	"Mecklenburg-Vorpommern" "Niedersachsen" "Nordrhein-Westfalen" "Rheinland-Pfalz" "Saarland" "Sachsen" "Sachsen-Anhalt" "Schleswig-Holstein" "Thüringen"
Promotion	Binary indicator for whether a promotion occurred (0 for no and 1 for yes)
Training Participation	Binary indicator for training programme participation (0 for no and 1 for yes)
Union Membership	Binary indicator for union affiliation
Parental Status	Binary indicator of whether the individual has dependents (0 for no and 1 for yes)
Wage Gap Indicator	A derived binary indicator showing if the wage is below the median for the individual's demographic group (gender, occupation type, and education level). For each subject, compare their annual wage to the median wage of their demographic group for that year. Then, if the individual's wage is below the median, assign a value of 1 (indicating a wage gap), and if the individual's wage is at or above the median, assign a value of 0

Throughout this entire analytical process, all of the coding is written in the SAS programming language, with PROC GLM and PROC LOGISTIC functioning as the primary computational engines for our analyses. Additionally, we include annotated starter scripts, allowing readers the opportunity to replicate every transformation and estimation process, from the initial raw data import to the final policy-ready interpretation of our findings. By the end of Chapter 5, readers will have fitted and interpreted both linear and logistic FE models using SAS Studio. These estimations, generated via PROC GLM and PROC LOGISTIC… STRATA statements reveal that, after purging all time-invariant subject-level traits, age emerges as the most potent predictor of within-subject wage variation. Meanwhile, education qualifications, particularly Bachelor's and Master's degrees, offer the clearest buffer against income disadvantage, even when promotions, sectoral shifts, and regional differences are accounted for.

These findings carry significant implications for wage equity policy settings, which we will examine in detail. By identifying these policy avenues, we effectively transform what could potentially be perceived as an abstract data-engineering task into a compelling value proposition. Each variable is defined, every panel diagnostic thoroughly executed, and every metadata tag carefully attached will ultimately serve to enhance the robustness of the evidence base that supports reforms in wage-related pay scales, the modernisation of collective agreements, and the reinforcement of Germany's statutory wage-transparency framework.

Case Study 2 – Understanding Education-Related Inequalities and Mortality

Case Study 2 extends this book's empirical arc from labour-market outcomes to health equity by examining how educational attainment shapes the risk of dying in mid-life. Drawing on the Person-Level Integrated Data Asset (PLIDA) data-generating recipes outlined in Chapter 2, the case study assembles a synthetic, yet similar distribution panel of 5,000 Australians observed annually over 5 years. Results reviewed in Chapter 6 foreshadow two seemingly contradictory insights: on the one hand, formal qualifications explain only a modest share of residual variation after full adjustment; on the other hand, the absolute gap in death risk between education groups accelerates sharply from the mid-forties onward. These twin findings frame this chapter's central objective, showing how advanced longitudinal models can surface small but actionable policy levers that would be invisible in cross-sectional snapshots.

The analytic file is a person-year panel, unbalanced by observation year, keyed by an artificial identifier (ID). Each record includes a one-year mortality indicator, exact age (20–99), educational attainment (High School, Bachelor's, Master's, Other Certificate), quintile household income, smoking status, indicators of chronic illness, healthcare access, and physical activity, and contextual markers such as region of residence. Because subjects can migrate, enrol late, or die, follow-up periods naturally differ, mirroring the incomplete trajectories that policymakers confront in real PLIDA extracts. The synthetic generator preserves realistic joint distributions among age, education, and mortality while guaranteeing confidentiality under Australia's Five Safes Framework.

Methodologically, this chapter progresses from a random-intercept logit, which accommodates time-invariant, subject-specific heterogeneity, to a random age-slope specification that allows ageing to vary across subjects, and finally to a model that interacts with education with calendar year to test whether inequality trends shift over time. Estimation relies on PROC GLIMMIX with Laplace integration and ridge-stabilised Newton–Raphson optimisation; predicted logits are converted to death probabilities by scoring the model on a fully crossed age-by-education grid in PROC PLM. A life-table algorithm then cumulates survival probabilities to determine the remaining life expectancy at 40, and years of life lost (YLL) are calculated relative to a benchmark for a Master's degree. Merging these YLL figures with cohort counts and stylised intervention costs yields incremental cost-effectiveness ratios (ICERs) that are directly comparable with Australia's AUD50,000 per life-year willingness-to-pay threshold.

TABLE 3.2
Case Study 2 Dataset Structure

Variable Name	Description
ID	Unique identifier for each subject
Year	The year of observation
Mortality Status	Binary outcome, alive/deceased by the end of the study period (1 for deceased)
Gender	Categorical (Male/Female), allowing for gender wage gap analysis
Age	Years (continuous)
Job Tenure	Years with the current employer (continuous)
Educational Level	Levels: "High School" "Master's" "Bachelor's" "Other Cert"
Income Quintile	Categorical variable indicating the income range (e.g., lowest quintile to highest quintile)
Employed	Binary Employment Status (1 for employed and 0 for unemployed)
Chronic Illness	Binary indicator for chronic health conditions (1 for its existence)
Access to Healthcare	Binary indicator for regular healthcare access (1 for its existence)
Physical Activity Level	Levels: "High" "Medium" "Low"
Region	Levels: "Urban" "Rural"
Smoking Status	Levels: "Never" "Current" "Former"
Migration Status	Binary variable (1 if the individual moved between regions during the study period)

From a policy perspective, the modelling paints a nuanced picture. Universal, low-intensity health-literacy campaigns aimed at adults aged 40 and above emerge as clear best buys, delivering additional life-years at roughly AUD3,000. Vocational training vouchers appear less attractive at face value (approximately AUD75,000 per life-year) but could enter the cost-effective zone if retargeted to health-enhancing occupations or bundled with mandatory "Healthy-Work" micro-credentials. Equity scholarships for university study appear expensive on health grounds alone, yet remain defensible when broader productivity and civic dividends are factored into a social cost-benefit analysis. The age gradient in absolute risk implies that proportionate universalism, offering the same interventions to everyone but intensifying support for socio-economically disadvantaged older adults, would maximise marginal health returns without widening inequities.

In short, by weaving RE modelling, life-table logic and cost analysis into a single narrative, this chapter demonstrates how modern longitudinal methods can move Australian health and education portfolios from statistical insight to evidence-based action.

Case Study 3 – Understanding Intergenerational Income Mobility

Intergenerational income mobility refers to the degree to which individuals are able to surpass their parents' economic status. High mobility implies that social and economic structures allow individuals to achieve economic outcomes independent of their background, while low mobility signals persistent inequality.

In this third case study, we turn to the question of income mobility using a synthetic panel dataset designed to mirror the structure and statistical properties of the Longitudinal Administrative Databank (LAD). This dataset follows 5,000 individuals over a five-year window and includes rich information on income, parental background, education, employment, and region, making it ideally suited for estimating marginal effects at the population level.

The statistical framework underpinning this case study is the GEE approach, introduced in Chapter 7. Unlike FE or RE models, which estimate subject-specific parameters and eliminate time-invariant covariates, GEE produces population-average estimates that retain all predictors, including static ones like parental income quintile. This is particularly important for mobility research, where variables such as family background and parental education are central to the theoretical and policy conversation but would be differenced out in an FE framework. By deploying GEE with both Gaussian (for log income) and binomial (for upward mobility) link functions, this case study offers two complementary lenses on mobility: (i) how subject income levels vary with parental background and personal attributes and (ii) what predicts the probability of achieving upward income movement relative to one's origin quintile. Each outcome is modelled using multiple correlation structures: independence, exchangeable, and first-order autoregressive to determine which best captures the longitudinal.

The empirical work, detailed in Chapter 7, demonstrates that education and employment status are the most powerful and consistent predictors of both absolute income levels and upward mobility odds. Subjects with only high school qualifications earn less than their tertiary-educated peers and are significantly less likely to exceed their parents' income position. Similarly, being employed is a strong positive predictor of both income and mobility. In contrast, parental income quintile predicts upward mobility but not income, once controls are added; and regional and year effects are largely insignificant, suggesting that geographic and temporal context play a smaller role in shaping economic trajectories than individual-level structural factors.

The modelling decisions made in this case study are grounded in the methodological and policy goals discussed in earlier chapters. First, GEE retains all subjects, including those with flat mobility status, ensuring representativeness of the population. Second, it allows us to compute robust

TABLE 3.3

Case Study 3 Dataset Structure

Variable Name	Description
ID	Unique identifier for each subject
Year	The year of observation
Income	Annual income (continuous variable)
Parental Income Quintile	Categorical variable indicating the income quintile of parents (e.g., lowest to highest)
Education Level	Levels: "High School" "Tertiary"
Employment Status	Levels: "Full-time" "Part-time" "Unemployed"
Region	Levels: "Quebec" "Saskatchewan" "Nova Scotia" "Prince Edward Island" "New Brunswick" "Ontario" "British Columbia" "Manitoba" "Newfoundland and Labrador" "Alberta"
Age	Years (continuous)
Gender	Categorical (Male/Female)
Household Size	Continuous
Occupation Type	Levels: "Administrative" "Managerial" "Technical"
Migration Status	Binary variable indicating whether the individual moved between regions (1 for moved between regions)
Parental Education Level	Levels: "University" "College" "High School"
Industry Sector	Levels: "Manufacturing" "Public" "Services"
Economic Shocks Indicator	Binary variable capturing major economic events, e.g., recession years (1 for shock)

standard errors even when the correlation structure is misspecified, making it an excellent candidate for applied policy work where data may not conform neatly to textbook assumptions. Third, the Quasi-likelihood Information Criterion values reported in Chapter 7 enable a rigorous, empirical selection of the best-fitting working correlation structure, ensuring that model selection is grounded in quasi-likelihood efficiency, not convenience.

Overall, this case study completes this book's methodological arc by showing how population-averaged models can produce actionable estimates in applied research. The GEE framework allows for principled inference about average effects, preserves time-invariant predictors, and handles attrition in unbalanced panel features that are especially valuable when translating data into policy. Readers will thus come away with a toolkit that spans from synthetic data construction to policy-facing statistical reporting, closing the loop between theory, implementation, and real-world application. From a policy perspective, this case study confirms the hypothesis that intergenerational mobility is governed less by structural geography or macroeconomic trends, and more by access to education and the labour market. This reinforces the value of investing in tertiary education pathways and employment facilitation strategies, especially for youth from disadvantaged backgrounds. It also provides empirical justification for shifting policy focus from background-targeted welfare to capability-enhancing public investment.

4

Exploratory Analysis of Longitudinal Data: Quality Audits, Visualisation, and Pre-Modelling Decisions

By the end of this chapter, readers will have acquired the essential data exploratory toolkit that serves as the foundation for every modelling step in this book. In particular, readers will understand how to implement a reproducible exploratory data analysis (EDA) checklist on any long-format panel: counting individuals and waves, analysing missingness, and comparing within- versus between-subject variations. Building on the data quality dimensions introduced in Chapter 1, readers will discover how to associate each diagnostic (completeness, consistency, uniqueness, accuracy), generate high-quality visualisations, such as histograms, trajectory plots, and heat maps, directly in SAS to illustrate the dynamics of subjects and groups with a specific SAS procedure and interpret the results for preparedness for modelling. Lastly, readers will practice transforming empirical patterns identified during the particular EDA, such as right-skewed wage distributions and limited within variance for specific predictors, into testable hypotheses and pre-modelling decisions to guide the fixed-effects (FE) specifications analysed in subsequent chapters.

EDA, within the context of population research, primarily aims to identify patterns, trends, and anomalies in datasets, thereby enhancing data comprehension and facilitating the formulation of novel hypotheses. It transcends being a mere collection of techniques; it embodies a methodological framework that integrates statistical modelling and graphical representation to unveil insights that may elude conventional confirmatory data analysis techniques. Therefore, EDA's central objective is to bolster and corroborate statistical conclusions while generating new hypotheses informed by observed data patterns. EDA proves particularly advantageous in population research as it empowers scholars to investigate data without pre-existing hypotheses, thereby facilitating the identification of unanticipated patterns and interrelations. It is an essential framework where data frequently manifests as intricate and multifaceted, necessitating a versatile and open-ended analytical approach to extract insights.

EDA is instrumental in helping researchers identify patterns and irregularities within their datasets by thoroughly examining the dataset's architecture and attributes. EDA typically represents the initial phase in the data

DOI: 10.1201/9781003599869-4

analysis pipeline, enabling researchers to validate data quality and extract insights from the dataset using descriptive statistics and visual representations such as histograms and boxplots. Moreover, EDA facilitates the identification of outliers and anomalies, which is crucial for understanding process inefficiencies and ensuring the validity of statistical premises. Using methodologies such as principal component analysis (PCA), EDA facilitates the unsupervised differentiation of samples, thereby enhancing the understanding of multivariate datasets.

A hallmark of EDA is the utilisation of interactive graphical tools, which augment the capacity for dynamic and iterative data exploration. EDA also encompasses data profiling, which evaluates data quality, alongside discovery, which aims to unveil new insights fundamental to the analytical framework. Data profiling is an approach to data quality analysis, and data quality is the mechanism for creating, collecting, storing, maintaining, transferring, processing, and presenting data to support business processes in a timely and cost-effective manner. This requires an understanding of the characteristics of the data that determine its quality and ability to measure, manage, and report on data quality. Data quality is a multifaceted concept shaped by six interrelated dimensions. "Completeness" gauges how much of the theoretically available information is captured, while "uniqueness" ensures no entity is recorded more than once under its identifying attributes. "Timeliness" reflects the degree to which a dataset mirrors reality at the required moment, and "validity" assesses whether each value conforms to its prescribed format, type, or range. "Accuracy" measures how faithfully the data describe real-world objects or events, and consistency demands that multiple representations of the same thing align with one another and with their underlying definition. Together, these dimensions provide a holistic framework for evaluating and improving any data asset's reliability, relevance, and utility.

Table 4.1 translates the six abstract data quality dimensions into concrete, reproducible SAS checks. Each row shows (i) the typical longitudinal problem, (ii) a one-line procedure or macro that generates the diagnostic, and (iii) the actionable output. Together, the commands form a minimalist yet comprehensive quality-assurance harness: PROC FREQ audits completeness by enumerating missing codes, PROC SORT … DUPOUT=flags duplication, range and cross-field rules expose inconsistency, format look-ups validate categorical codes, joins to external benchmarks test accuracy, and lag summaries quantify timeliness.

Running and saving the complete set of diagnostics satisfies the transparency principles stressed in Chapter 2 and equips the readers with an annotated quality record that downstream analysts and reviewers can trace when they interpret the FE and RE results.

Data profiling is the process of analysing and assessing data quality across data quality dimensions, and metadata profiling involves assessing and examining the data structures. The results can be used to evaluate compliance with enterprise-wide standards and content profiling, including

TABLE 4.1

Data Quality Dimensions and SAS Checks

Data Quality Dimensions	Typical Longitudinal Issue	SAS Diagnostics	Output
Completeness (missingness)	Linkage gaps; item non-response	PROC FREQ Data=panel; tables _all_ / missing; run; —or— a %MISS_REPORT macro	Cell counts of "." or blank by variable and wave; feeds the missing-value heat-map you plot later
Uniqueness (duplicate IDs/records)	Same person/record entered twice	PROC SORT Data=panel NODUPKEY; by ID Year; run; with DUPOUT=Option to capture duplicates	List of duplicate rows to inspect or delete
Consistency (cross-field coherence)	Age decreases over time; negative wages	(i) PROC COMPARE Base=Panel Compare=Panel noprint out=agechk; to detect age back-sliding (ii) simple range checks in a %RULE_ENGINE macro	Table of rule violations (e.g., Age_t<Age_{t-1}) marked for correction
Validity (conformance to allowed values)	Invalid occupation codes; out-of-range dates	PROC FORMAT Cntlout=fmt;+PROC FREQ; tables OccupationCode/missing; to cross-tab with format library	Frequency of codes not found in the official job classification
Accuracy (agreement with trusted source)	Wage outside legal minimum; Age inconsistent with education level	Merge to external control totals (e.g., minimum wage table) and run PROC SQL exceptions query	List of observations violating external benchmarks for follow-up
Timeliness/Currency	Lag between event date and data-entry date	PROC MEANS Data=Panel n mean max; var EntryLag; class Year; where EntryLag=EntryDate – EventDate	Summary of reporting lags by year, large lags may bias recent waves

determining and reviewing the data content. The results of this assessment reflect the quality of the source data's content and identify issues that will be resolved through data transformations.

The following sections of this chapter will use the data from Case Study 1. Using SAS Studio, the aim is to identify patterns, trends, and anomalies within the dataset, thereby augmenting our understanding before formal statistical modelling design. The principal instruments we will deploy to analyse the dataset include SAS PROC MEANS, SGPLOT, FREQ, PRINT, and FACTOR. The SAS code will be provided.

Conducting EDA for a particular longitudinal dataset necessitates visualising as much of the relevant raw data as feasible, accentuating aggregate patterns of potential significance, recognising both cross-sectional and longitudinal trends, and pinpointing atypical observations.

A scatter plot illustrating the response over time is a valuable visualisation. Connecting the repeated measurements for each subject across time elucidates whether a recognisable pattern is prevalent among the subjects. These subject profiles may also yield insights into between-subject variability. Visualising observed profiles over time aids in discerning general trends within subjects, may uncover non-linear changes throughout time, and offers insights into the variability at specific time points.

To analyse the data presented in Case Study 1, it is often helpful to generate a SAS line listing report that details each data point while simultaneously calculating and presenting descriptive statistics that elucidate the characteristics of the numeric variables contained within the data. In addition to these statistical representations, creating a series of visual graphs that illustrate both the individual profiles of the subjects and the collective group profiles, thereby facilitating a deeper understanding of the underlying patterns and relationships present in the dataset, will be helpful. This multifaceted approach enhances our analytical capabilities and provides a robust framework for interpreting the findings.

For instance, Figure 4.1 presents a graph depicting wage variations among multiple subjects. These distinct profiles reveal several significant trends: some subjects experienced wage increases, while others encountered

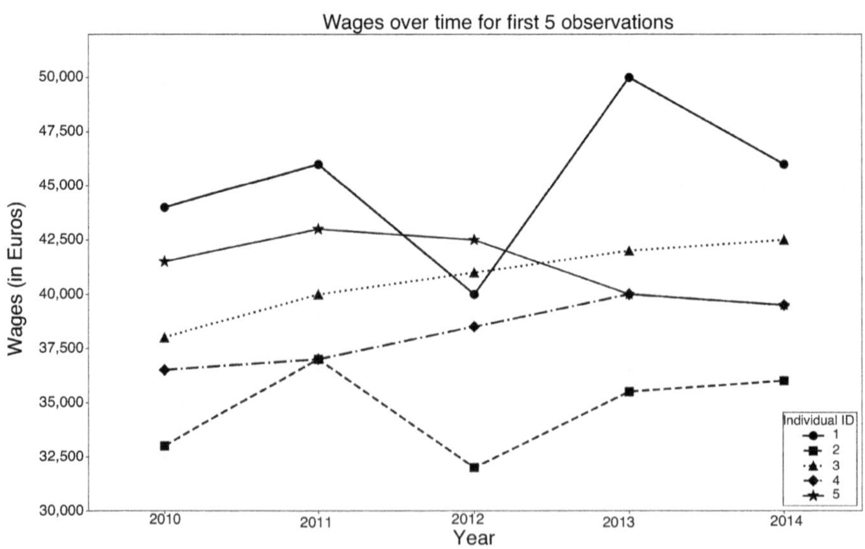

FIGURE 4.1
Temporal data visualisation.

wage declines. Subjects who commenced the study with the highest wages generally maintained their status as the top earners throughout the investigation, and the measurement variability is notably less pronounced among those earning lower wages. In addition to charting the responses over time, it is also helpful to incorporate various subgroups within the same graph to elucidate the correlation between wages over time. For instance, male and female subjects have experienced wage growth as shown in Figure 4.2; however, the decline in male subjects' wage growth appears more substantial than that of female subjects, suggesting a potential interaction between gender and time.

Let us analyse the data by generating a comprehensive SAS line listing report and descriptive statistics pertinent to the numeric variables. Later, we will construct graphical representations that illustrate both subjects and group profiles. The PROC PRINT procedure will yield the tabular representation depicted in Table 4.1. As is noticed, age and gender are considered time-invariant variables; however, gender may also be interpreted as time-varying. All remaining variables are classified as time-varying variables. The dataset utilised in Case Study 1 exhibits balance, as the number of observations remains consistent across subjects. The binary variables, namely Parental_Status, Training_Participation, Union_Membership, and Wage_Gap_Indicator, are assigned a value of "0" in the absence of the corresponding event and a value of "1" otherwise. The SAS syntax will perform the listing presented in Table 4.1.

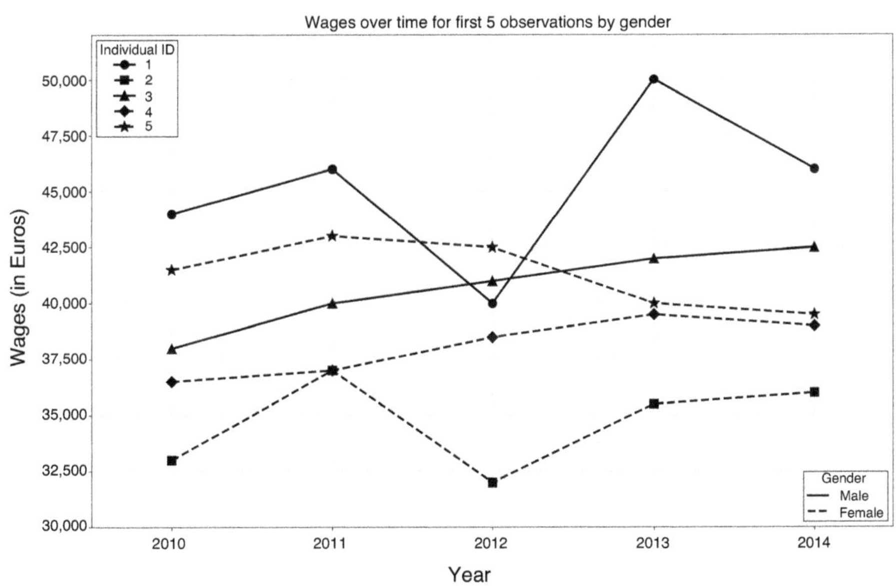

FIGURE 4.2
Temporal, grouped data visualisation.

SAS Syntax 4.1

```
proc print data=Case_Study_1 (obs=10) label;
  var ID Year Age Gender Parental_Status General_Education
      Vocational_Qualification Region Occupation Sector
      Firm_Size Job_Tenure Base_Wage Wage Median_Wage
      Training_Participation Union_Membership Promotion
      Wage_Gap_Indicator;
  title "Case-Study 1 – First 10 Observations (panel-long format)";
run;
```

What the Code Does

This snippet prints a labelled line listing of the first ten observations. By restricting the VAR list to the numeric variables defined in Chapter 3 and using the long-format panel file, it lets you confirm at a glance that (i) IDs repeat across years as expected, (ii) each variable has been imported under the correct name and retains its descriptive labelling, and (iii) the derived *Wage_Gap_Indicator* appears in the dataset. In other words, it is a visual sanity check that the data-import and reshaping pipeline has worked before we move on to more elaborate exploratory diagnostics.

The line listing of Case Study 1's dataset provides the reader with information on the nature of the variables that constitute it. As a very helpful SAS procedure, the PROC MEANS procedure will yield the tabular representation depicted in Table 4.2 that shows aggregate statistics on the Case Study 1's dataset, which contains 25,000 observations covering age, wages, and selected employment characteristics.

TABLE 4.2

Line Listing, First Ten Observations

ID	Year	Age	Gender	Region	...	Promotion	Wage_Gap_ Indicator
1	2010	31	0	Baden-Wurttemberg	...	0	0
2	2010	54	1	Baden-Wurttemberg	...	0	0
3	2010	33	0	Baden-Wurttemberg	...	1	1
4	2010	56	1	Nordrhein-Westfalen	...	0	1
5	2010	55	1	Mecklenburg-Vorpommern	...	0	0
6	2010	31	1	Bayern	...	0	0
7	2010	53	1	Nordrhein-Westfalen	...	0	1
8	2010	22	1	Nordrhein-Westfalen	...	0	1
9	2010	44	1	Bayern	...	1	0
10	2010	39	0	Hessen	...	1	1

The outcome variable, *Wage*, ranges from EUR22,385 to EUR67,463, whereas the *Age* variable ranges from 18 to 69, with a mean of approximately 43.3 years (SD ≈ 14). Wages show considerable variation, averaging about EUR38,638.85 (SD ≈ 7,802). The variables *Training_Participation, Year_of_Promotion, Union_Membership,* and *Wage_Gap_Indicator* are binary variables, where 1 indicates that the subject participated in a training programme, is a union member, and the subject's wage is below the median in the respective year. The mean of 0.2 for the *Training_Participation* variable indicates that 20 per cent of the observations reported participation in training programmes, and the mean of 0.32 for the *Union_Membership* variable indicates that 31 per cent of the observations reported participation in union membership. The mean of 0.47 for the *Wage_Gap_Indicator* variable indicates that 47 per cent of the observations had instances where the subject's wage was below the median.

Here is the SAS syntax that will perform this analysis:

SAS Syntax 4.2

```
proc means data=Case_Study_1 n min max mean median std maxdec=2;
   var  Age Wage Job_Tenure
        Training_Participation Union_Membership Wage_Gap_Indicator;
   title "Case-Study 1 – Descriptive Statistics (panel-long file)";
run;
```

What the Code Does

This syntax produces a compact numerical profile of the dataset's main variables, reporting the count of non-missing observations (*n*), minimum, maximum, arithmetic mean, median, and standard deviation (rounded to two decimals). Running it at the start of an exploratory workflow serves two purposes: it confirms that each variable has plausible ranges and provides baseline dispersion measures that you will later compare with the within- and between-subject standard deviations in the FE diagnostics that will be explored in Chapter 5. The NOROW, NOCOL, and NOPCT options can suppress the percentage calculations so that only the raw frequency counts appear in Table 4.3.

Analysing the distribution of numeric data using histograms and kernel density estimators is essential for understanding the data's underlying probability density function. Histograms provide a straightforward, albeit coarse, representation of data distribution by dividing the data into bins and counting the number of observations in each bin. This method is beneficial for its simplicity and ease of interpretation, making it a common choice for initial EDA. The following SAS syntax generates the distribution of the *Wage* variable, displayed as a histogram overlay with a kernel density function.

TABLE 4.3

Selected Descriptive Statistics

Variable	N	Min	Max	Mean	Median	Std Dev
Age	25,000	18	69	43.26	43	13.91
Wage	25,000	4,469.20	72,627.70	38,638.85	38,667.40	7,802.03
Job_Tenure	25,000	1	5	3	3	1.41
Training_ Participation	25,000	0	1	0.2	0	0.4
Union_ Membership	25,000	0	1	0.31	0	0.46
Wage_Gap_ Indicator	25,000	0	1	0.47	0	0.5

SAS Syntax 4.3

```
title "Wage Distribution with Kernel Density Function";
proc sgplot data=Case_Study_1;
   histogram Wage/nbins=30 transparency=0.2;
   density  Wage/type=kernel lineattrs=(color=red thickness=2);
   xaxis label="Hourly Wage (€)" grid;
   yaxis label="Frequency"    grid;
run;
title;
```

What the Code Does

This PROC SGPLOT call visualises the empirical wage distribution by combining a 30-bin histogram (semi-transparent bars) with a non-parametric kernel density curve. The display quickly reveals skewness, potential outliers, and multi-modality, which is the information that guides modelling decisions such as log-transforming wages or trimming extreme values. Gridlines and clear axis labels make the graphic publication-ready and consistent with the visual style adopted elsewhere in the EDA chapter.

Figure 4.3 illustrates the histogram analysis alongside the superimposed kernel density estimation for the *Wage* variable. The overlaid kernel density function on the histogram provides a refined estimation of the probability density function, smoothing the data distribution's representation. This juxtaposition clarifies wage distribution patterns, revealing a near-normal distribution centred on the mean wage, while enhancing comprehension through a continuous lens. As is seen from the histogram, the distribution exhibits a unimodal characteristic (single peak), with the highest concentration observed in the EUR40,000–EUR45,000 interval. The kernel density curve (depicted in bold line) indicates a relatively smooth, bell-shaped form, although there is evidence of a right skew (an extended tail on the higher-wage

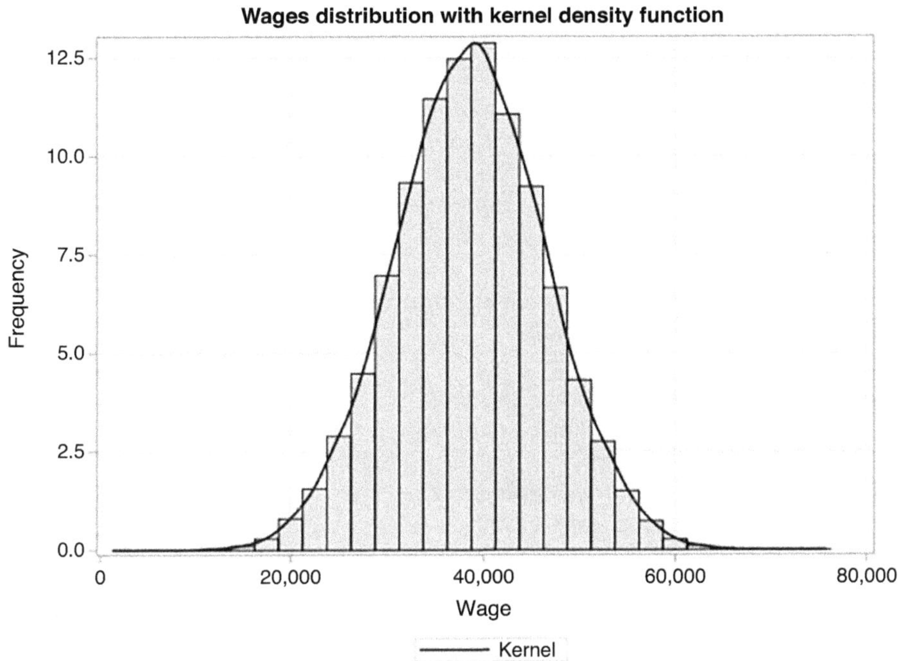

FIGURE 4.3
Histogram, wages.

end). Most wages are distributed over a EUR30,000–EUR55,000 interval, with the mode occurring near the mid-40k region. A limited number of observations surpass EUR60,000, reaching approximately EUR80,000; yet, the frequencies within that segment are comparatively low. The histogram, on the other hand, illustrates that while most subjects earn within a moderate range of around EUR40,000–EUR50,000, there remains a discernible tail extending toward elevated wages, indicative of a marginally skewed distribution.

Now, let us examine categorical variables in conjunction with the wage. The following discussion illustrates wage trends by occupation, age, gender, region, and education. Readers are invited to utilise the conceptual examples and the SAS syntax presented to implement these methodologies in their research. Let us examine the evolution of wages over the years, segmented by age demographics. The subsequent array of SAS codes generates line graphs, facilitating the identification of patterns and anomalies, thereby enhancing our understanding.

SAS Syntax 4.4

```
data Case_Study_1_binned;
   set Case_Study_1;
```

```
    length Age_Group $8;            /* shortest label length = '55+'   */
    if    Age  < 25 then Age_Group = 'Under 25';
    else if Age < 35 then Age_Group = '25–34';
    else if Age < 45 then Age_Group = '35–44';
    else if Age < 55 then Age_Group = '45–54';
    else                 Age_Group = '55+'; run;
proc means data=Case_Study_1_binned nway noprint;
    class Year Age_Group;
    var  Wage;
    output out=MeanWageByAgeGroup(drop=_type_ _freq_)
        mean = Mean_Wage; run;
title "Hourly-wage trend by age group";
proc sgplot data=MeanWageByAgeGroup;
    series x=Year  y=Mean_Wage/group=Age_Group
                    markers
                    lineattrs=(thickness=2);
    xaxis label="Calendar year" grid;
    yaxis label="Average hourly wage (€)" grid;
run;
title;
```

What the Code Does

This three-step routine first enriches the panel by transforming the continuous Age variable into a five-level categorical *Age_Group*, making intra-cohort wage patterns easier to visualise. It then collapses the data with PROC MEANS, computing the mean hourly wage for every unique *Year* × *Age_Group* combination and writing the result to a summary table. Finally, PROC SGPLOT renders those means as a set of colour-coded lines, one per age band, allowing you to see how the age premium evolves across calendar years. The figure provides an immediate, visual check on whether wage growth is uniform across cohorts. This cue later motivates the *Age* and *Age* × *Year* coefficients estimated in Chapter 5's FE models.

As shown in Figure 4.4, from 2010 to 2014, all age groups exhibited a consistent average wage increase, indicating a consistent upward wage trend. The 55+ age group consistently displays the highest average wages for most of this duration, signifying that older workers typically earn the most. The under-25 demographic begins as, and consistently remains, one of the lower-wage tiers, a characteristic of entry-level or nascent career professionals. The 25–34 cohort starts at a relatively low wage level but exhibits a markedly steep ascent by 2014, diminishing the disparity with older age demographics. By 2013–2014, convergence was observable, yet the 55+ group still prevailed, and the under-25 years cohort remained disadvantaged. This analysis indicates wage advancement across all age groups, albeit with enduring age-related distinctions.

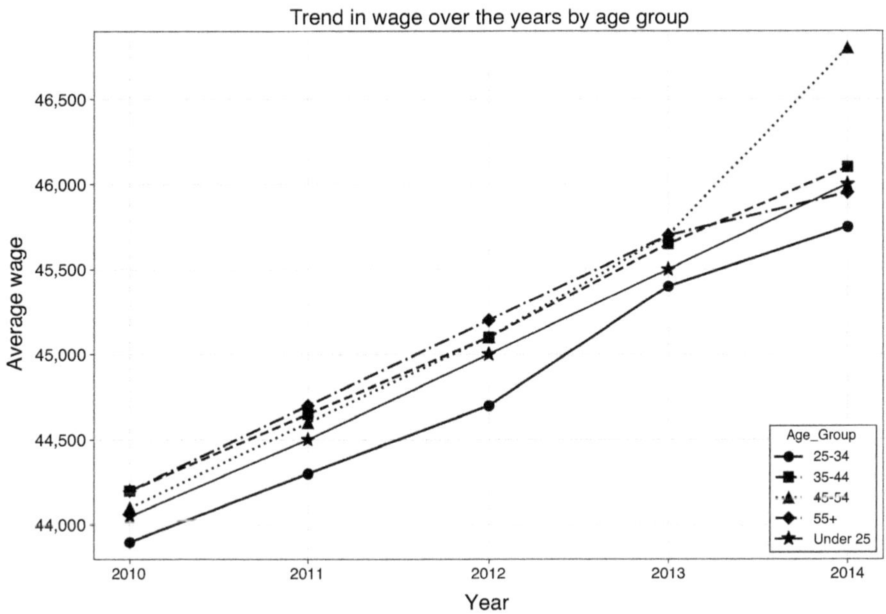

FIGURE 4.4
Wage trends by age groups line chart, 2010–2014.

Let us now create a line chart depicting wage trends over the years by gender with the following SAS syntax.

SAS Syntax 4.5

```
proc means data=Case_Study_1 nway noprint;
   class Year Gender;
   var  Wage;
   output out=MeanWageByGender(drop=_type_ _freq_)
        mean = Mean_Wage; run;
title "Hourly-wage trend by gender";
proc sgplot data=MeanWageByGender;
   series x=Year y=Mean_Wage/group=Gender
                    markers
                    lineattrs=(thickness=2);
   xaxis label="Calendar year" grid;
   yaxis label="Average hourly wage (€)" grid;
run;
title;
```

What the Code Does

This code calculates the mean hourly wage for every unique combination of *Year* and *Gender* and stores the results in a compact summary table. It then visualises those means as two separate lines, one for men, one for women, so readers can see whether the gender pay gap is widening, narrowing, or holding steady across the calendar years covered in the panel. The graph provides an immediate diagnostic of trend symmetry (or lack thereof), information that feeds directly into later modelling decisions such as including a Year main effect, a *Gender* × *Year* interaction, or a gender-specific FE specification.

Figure 4.5, depicting Wage Trends by Gender, illustrates that both categories (Gender=0 and Gender=1) exhibit a consistent upward trend in average wages from 2010 to 2014, indicating a reliable pattern of wage advancement over the specified timeframe. Nevertheless, throughout this duration, the cohort denoted as "0", that is, "Male", consistently shows higher average wages than the cohort designated as "1", that is, "Female".

This phenomenon highlights an enduring disparity between the two groups, suggesting that although the remuneration for all individuals appears to rise, males consistently earn more than females, and the gap does not diminish significantly over the years under investigation.

Similarly, the line chart analysis depicting the trend of wages by occupation over the years, as shown in Figure 4.6, reveals an analogous upward trajectory similar to that observed across various age demographics and genders. Each occupational category in this chart shows a consistent increase in

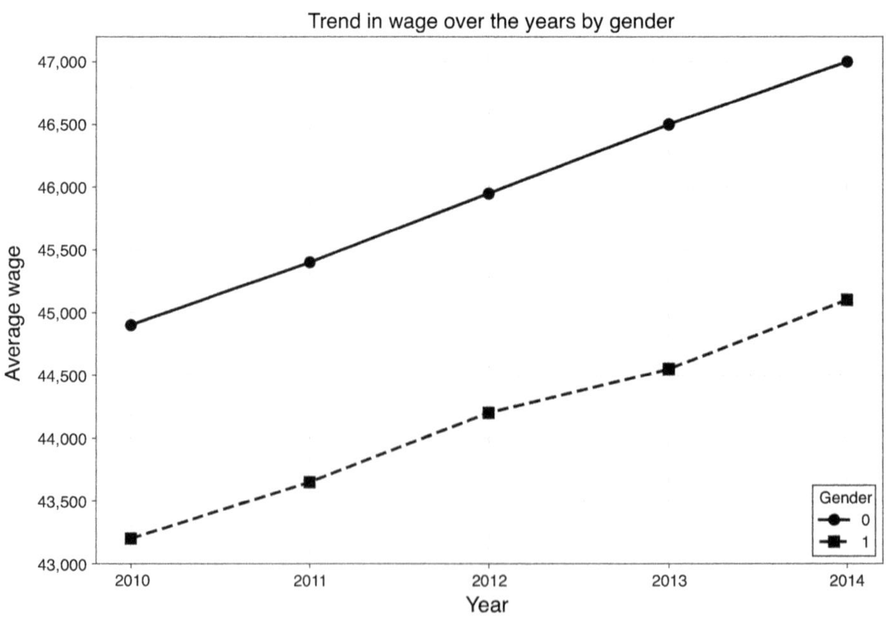

FIGURE 4.5
Wage trends by gender line chart, 2010–2014.

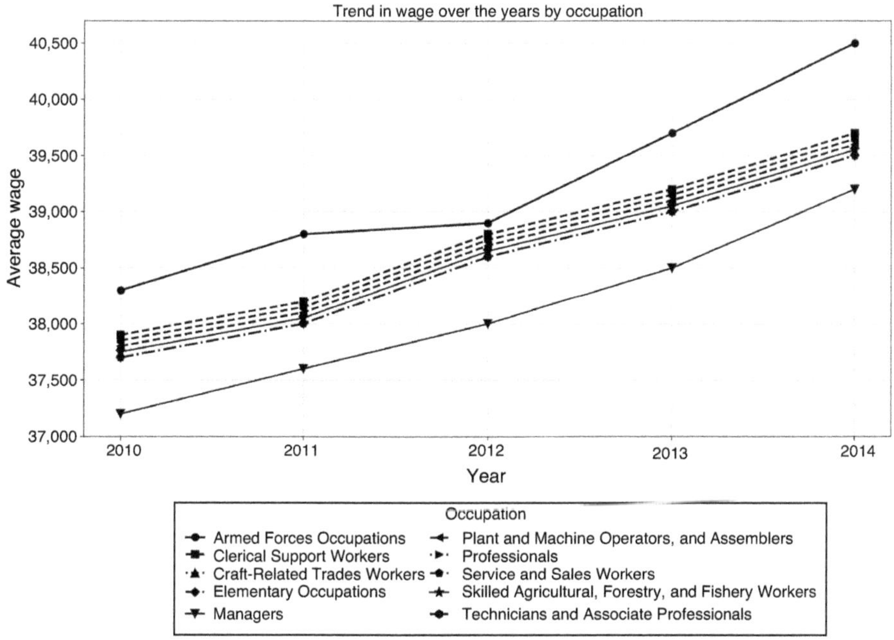

FIGURE 4.6
Wage trends by occupation line chart, 2010–2014.

average wages from 2010 to 2014. Although all groups exhibit wage growth, significant disparities in initial and final wage levels are evident; certain occupations start above the overall average and maintain their advantage, whereas others begin at a lower level yet still experience substantial growth over time. By 2014, these occupational trajectories had diverged, indicating enduring wage-level disparities across various professions, despite a general upward trend. This observation aligns with the patterns identified in prior charts, suggesting that pre-existing disparities persist while wages increase uniformly across the spectrum. Next, we will investigate average wages by occupation by generating a line chart with the following SAS syntax.

SAS Syntax 4.6

```
proc means data=Case_Study_1 nway noprint;
   class Year Occupation;
   var  Wage;
   output out=MeanWageByOcc(drop=_type_ _freq_)
        mean = Mean_Wage;
run;
title "Hourly-wage trend by occupation";
```

```
proc sgplot data=MeanWageByOcc;
   series x=Year y=Mean_Wage/group=Occupation
                        markers
                        lineattrs=(thickness=2);
   xaxis label="Calendar year" grid;
   yaxis label="Average hourly wage (€)" grid;
run;
title;
```

What the Code Does

This two-step block first aggregates the panel, producing a compact table of average hourly wages for every *Year* × *Occupation* combination. Next, it uses PROC SGPLOT to draw a separate time-series line for each occupation group, with markers identifying annual means. The resulting figure lets you see which occupations lead or lag in pay growth over the observation window, information that can prompt questions about sectoral bargaining agreements, skill shortages, or automation pressures and will later inform the choice of occupation dummies and interaction terms.

The examination of the temporal evolution of wages by region, as illustrated in Figure 4.7, reveals a pattern analogous to those identified across demographic

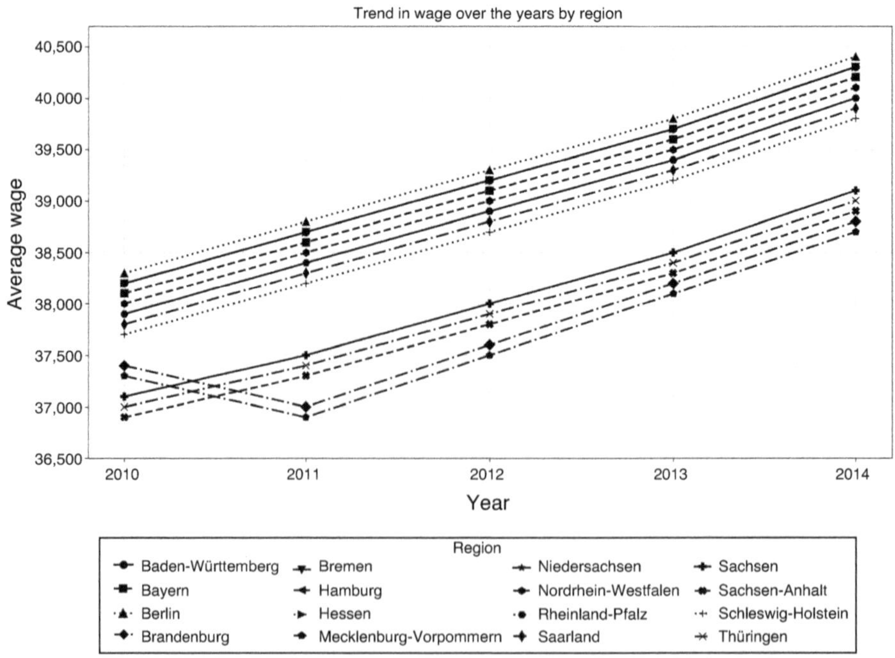

FIGURE 4.7
Wage trends by region line chart, 2010–2014.

variables such as age, gender, and occupation. Specifically, all regions exhibit a consistent increase in average wage levels from 2010 to 2014, signifying a widespread upward trajectory. Nevertheless, certain regions initiate at comparatively elevated average wage levels and sustain their advantage, whereas others persist at lower wage benchmarks, thereby highlighting enduring regional disparities. While it is evident that wages in every region experience growth over the specified period, the trajectories do not fully align, indicating that existing wage inequalities are maintained despite the overall increase in wage levels.

SAS Syntax 4.7

```
proc means data=Case_Study_1 nway noprint;
   class Year Region;
   var  Wage;
   output out=MeanWageByRegion(drop=_type_ _freq_)
         mean = Mean_Wage; run;
title "Hourly-wage trend by region";
proc sgplot data=MeanWageByRegion;
   series x=Year y=Mean_Wage/group=Region
                    markers
                    lineattrs=(thickness=2);
   xaxis label="Calendar year" grid;
   yaxis label="Average hourly wage (€)" grid;
run;
title;
```

What the Code Does

This code block first collapses the panel to yearly regional averages, giving a table in which each row represents a unique combination of *Year* and *Region* alongside its mean wage. It overlays those means as separate time-series lines, so readers can visually compare wage trajectories across regions. Steeper or flatter slopes flag regions whose earnings growth diverges from the national pattern, guiding later modelling choices (e.g., whether to include *Region dummies* or *Region* × *Year* interactions) and providing empirical context for the policy discussion of spatial wage disparities that follows in Chapters 5 and 6.

Case Study 1's dataset has two educational variables, *General_Education* and *Vocational_Qualification*. By employing advanced techniques to represent and elucidate the intricate interplay between these variables visually, readers may derive nuanced insights into the complex relationship between the level of educational attainment achieved by subjects and their corresponding wages in the labour market. Like previous analyses, let us use PROC MEANS and

SGPLOT to generate trends in average wages by education level. Here is the SAS syntax to create the analysis output.

SAS Syntax 4.8

```
proc means data=Case_Study_1 nway noprint;
   class Year General_Education;
   var  Wage;
   output out=MeanWageByEduc(drop=_type_ _freq_)
        mean = Mean_Wage; run;
title "Hourly-wage trend by general-education level";
proc sgplot data=MeanWageByEduc;
   series x=Year y=Mean_Wage/group=General_Education
                  markers
                  lineattrs=(thickness=2);
   xaxis label="Calendar year" grid;
   yaxis label="Average hourly wage (€)" grid;
run;
title;
```

What the Code Does

This two-step syntax first collapses the panel to produce a tidy summary table containing the average hourly wage for every *Year × General_Education* combination. PROC SGPLOT then draws separate time-series lines, one for each education tier, so readers can see how earnings trajectories diverge across schooling levels. Visualising the data this way immediately reveals whether the wage premium for higher credentials is widening, narrowing, or stable over time, an insight that foreshadows Chapter 5's FE finding that bachelor's and master's degrees reliably insulate workers from below-median earnings.

Figure 4.8, illustrating the correlation between wages and general education levels, indicates that from 2010 to 2014, each educational classification experienced a wage increase, suggesting a pervasive upward trend regardless of educational attainment. Nevertheless, subjects possessing a University Entrance Qualification commence and sustain the highest average wage, highlighting the significant influence of advanced qualifications. Concurrently, individuals belonging to lower-credential categories (for instance, those without a Secondary School Certificate) consistently receive lower earnings despite also experiencing wage increments. The wage disparities associated with educational attainment remain evident, with a pronounced relationship between higher academic qualifications and increased remuneration throughout the analysed timeframe. While all vocational

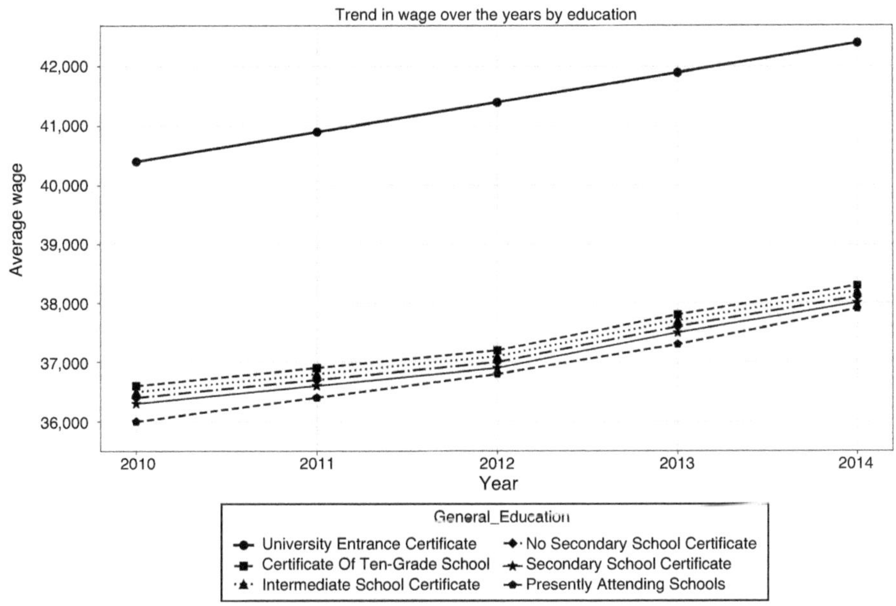

FIGURE 4.8
Wage trends by general education line chart, 2010–2014.

qualification categories exhibited consistent wage growth, individuals with a Doctorate degree consistently achieved the highest average wages, highlighting the benefits of advanced qualifications in earnings. In contrast, individuals with lower or no vocational qualifications earned less but experienced wage increases over time. This trend reaffirms the correlation between higher-level vocational qualifications and increased earnings despite wage growth across all qualification tiers. Repeating the analysis with the vocational education variable using the following SAS syntax provides a more nuanced understanding of the relationships between growth in average wages and vocational education completions.

SAS Syntax 4.9

```
proc means data=Case_Study_1 nway noprint;
   class Year Vocational_Qualification;
   var  Wage;
   output out=MeanWageByVoc(drop=_type_ _freq_)
        mean = Mean_Wage; run;
title "Hourly-wage trend by vocational qualification";
```

```
proc sgplot data=MeanWageByVoc;
  series x=Year y=Mean_Wage/group=Vocational_Qualification
                   markers
                   lineattrs=(thickness=2);
  xaxis label="Calendar year" grid;
  yaxis label="Average hourly wage (€)" grid;
run;
title;
```

What the Code Does

This code block first collapses the panel to obtain the average hourly wage for every *Year* × *Vocational Qualification* combination, writing the result to a compact table. It then plots those averages as distinct time-series lines, allowing you to track how earnings trajectories diverge among qualification tiers, apprenticeship, trade school, bachelor's degree, master's degree, etc. The figure provides an immediate visual cue for whether the education dividend is widening, narrowing, or steady across calendar time.

Upon the satisfactory completion of visualising observed profiles over time for a specific longitudinal dataset, which demands the visualisation of as much pertinent raw data as is practicable, population researchers would

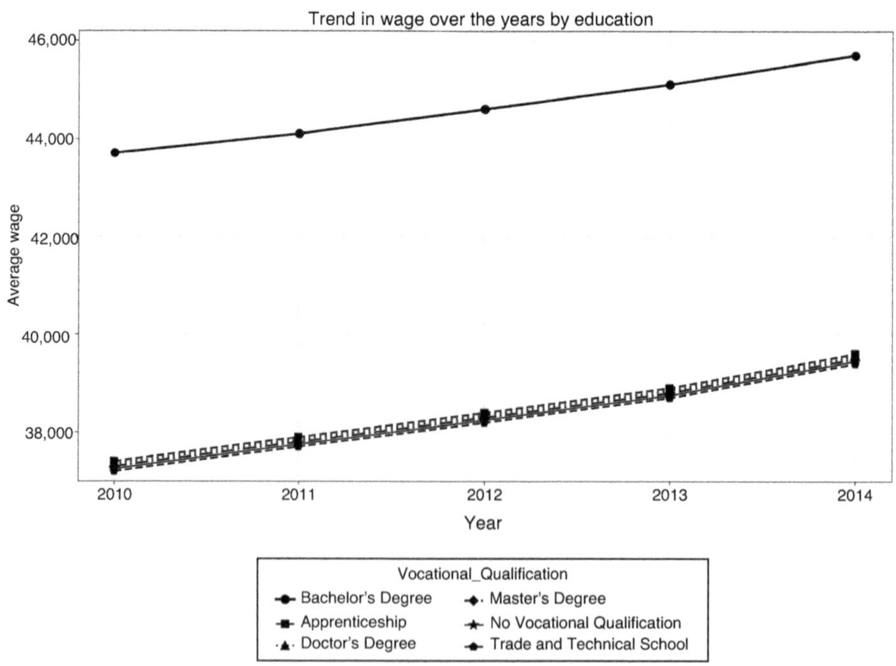

FIGURE 4.9
Wage trends by vocational qualification line chart, 2010–2014.

have synthesised patterns of potential scientific relevance, identified longitudinal trends, and highlighted atypical subjects or observations that are particularly aligned with their research objectives. The subsequent phase in EDA involves examining contingency tables, which can be classified as a relatively uncomplicated descriptive technique for investigating the relationship between a categorical variable (e.g., *Wage_Gap_Indicator*) and additional categorical variables (such as *Gender* and *Firm_Size*) using PROC FREQ in SAS. This methodological approach enables researchers to enumerate the frequency of each category combination, scrutinise proportions or percentages across various levels, and assess independence or association via a Chi-Squared test.

The subsequent SAS syntax facilitates the generation of contingency tables, specifically the *Wage_Gap_Indicator*, which serves as our dependent variable, alongside various categorical variables identified within the Case Study 1 dataset. Including the CHISQ option empowers the execution of Chi-Squared tests to assess associations. For instance, the FREQ procedure yields a cross-tabulation encompassing both row and column frequencies and table, row, and column percentages. In contrast, the CHISQ procedure produces a comprehensive array of Chi-Squared statistics, which includes degrees of freedom, test statistics, and their corresponding probabilities. The Chi-Squared statistical measures assess the null hypothesis, asserting the independence of the row and column variables within a cross-tabulation framework. Although the measures may suggest a relationship between two variables, they do not provide insights regarding the strength or directionality of this relationship.

SAS Syntax 4.10

```
proc freq data=Case_Study_1;
   tables Wage_Gap_Indicator * Gender/chisq norow nocol nopercent;
run;
```

What the Code Does

This PROC FREQ call cross-tabulates the binary *Wage Gap Indicator* against *Gender*, prints cell counts, and performs Pearson, likelihood ratio, and (if needed) Fisher's Exact Chi-Squared tests. At the exploratory stage, the code answers a policy-relevant question: "Within the pooled panel, are men and women equally likely to fall below their demographic-peer median wage?"

Let us now examine the table concerning the *Wage_Gap_Indicator* differentiated by Gender, which systematically cross-tabulates two binary variables: *Wage_Gap_Indicator* (0=absence of gap, 1=presence of gap) and *Gender*

(0=Male, 1=Female). The observed frequencies exhibit a relatively equitable distribution across the categories. From this analysis, it is evident that SAS has generated a 2×2 contingency table (*Wage_Gap_Indicator* by *Gender*) alongside various statistical measures to evaluate the presence of a relationship between the two categorical variables (Tables 4.4–4.6).

TABLE 4.4

FREQ Procedure

Table of Wage_Gap_Indicator by Gender				
		Gender		
Wage_Gap_Indicator		0	1	Total
Frequency	0	6,598	6,597	13,195
Per cent		26.39	26.39	52.78
Row Pct		50.00	50.00	
Col Pct		52.81	52.75	
Frequency	1	5,897	5,908	11,805
Per cent		23.59	23.63	47.22
Row Pct		49.95	50.05	
Col Pct		47.19	47.25	
	Total	12,495	12,505	25,000
		49.98	50.02	100.00

TABLE 4.5

Statistics for Table of Wage_Gap_Indicator by Gender

Statistic	DF	Value	Prob
Chi-Square	1	0.0063	0.9366
Likelihood Ratio Chi-Square	1	0.0063	0.9366
Continuity Adj. Chi-Square	1	0.0045	0.9467
Mantel-Haenszel Chi-Square	1	0.0063	0.9366
Phi Coefficient		0.0005	
Contingency Coefficient		0.0005	
Cramer's V		0.0005	

TABLE 4.6

Fisher's Exact Test

Cell (1,1) Frequency (F)	6598
Left-sided Pr≤F	0.5367
Right-sided Pr≥F	0.4733
Table Probability (P)	0.0101
Two-sided Pr≤P	0.9394

At first, the frequency table reveals that the data are almost evenly distributed across both categories of *Wage_Gap_Indicator* (0 vs. 1) within the two Gender classifications (coded as 0 vs. 1). The percentages for both rows and columns indicate slight variations in the frequency of wage gap status associated with either gender, suggesting that there may be a minimal or vwnon-existent association. The calculation of the Pearson Chi-Squared statistic reveals a value of 0.0063, alongside a p-value of 0.9366. This high p-value indicates that the sample lacks evidence of a statistically significant relationship between *Wage_Gap_Indicator* and *Gender.* Analysing the Likelihood Ratio Chi-Squared, the Continuity Adjusted Chi-Squared, and the Mantel-Haenszel Chi-Squared reveals a comparable conclusion, given that their p-values are very similar, thus reinforcing the lack of a relationship within the data. Moreover, the effect size measures (Phi Coefficient, Contingency Coefficient, and Cramer's V) are all remarkably low (0.0005), suggesting that, should a relationship exist, its scale would be trivial. Ultimately, Fisher's Exact Test, particularly relevant for 2×2 contingency tables, produces a non-significant two-sided p-value of roughly 0.9394, corroborating the outcomes from the Chi-Squared tests. Given the substantial sample size of 25,000, it is improbable that this null result can be attributed to insufficient statistical power; instead, it appears that there is indeed no significant correlation between *Wage_Gap_Indicator* and *Gender* within this dataset.

Furthermore, Table 4.7 presents a contingency table delineating the binary *Wage_Gap_Indicator* across multiple categories of *General_Education.* Each cell's frequency and percentage distributions elucidate the observed counts and their corresponding proportions across various education levels (Table 4.8).

TABLE 4.7

FREQ Procedure

Table of Wage_Gap_Indicator by General_Education

	General_Education						
	Certificate of Ten-Grade School	Intermediate School Certificate	No Secondary School Certificate	Presently Attending Schools	Secondary School Certificate	University Entrance Qualification	Total
Frequency 0	815	3,141	607	565	3,730	4,337	13,195
Per cent	3.26	12.56	2.43	2.26	14.92	17.35	52.78
Row Pct	6.18	23.80	4.60	4.28	28.27	32.87	
Col Pct	56.40	52.33	58.82	60.30	51.86	51.69	
Frequency 1	630	2,861	425	372	3,463	4,055	11,805
Per cent	2.52	11.44	1.70	1.49	13.85	4,054	47.22
Row Pct	5.34	24.24	3.60	3.15	29.34	16.22	
Col Pct	43.60	47.67	41.18	39.70	48.14	34.34	
Total	1,445	6,002	1,032	937	7,193	8,391	25,000
	5.78	24.01	4.13	3.75	28.77	33.56	100.00

TABLE 4.8

Statistics for Table of Wage_Gap_Indicator by General_Education

Statistic	DF	Value	Prob
Chi-Square	5	53.3349	<.0001
Likelihood Ratio Chi-Square	5	53.7054	<.0001
Mantel-Haenszel Chi-Square	1	8.9417	0.0028
Phi Coefficient		0.0462	
Contingency Coefficient		0.0461	
Cramer's V		0.0462	

In examining the inferential outcomes, the Pearson Chi-Squared (53.3349, df=5) yields a p-value<0.0001, signifying a statistically significant correlation between *Wage_Gap_Indicator* and *General_Education*. This conclusion is further substantiated by the Likelihood Ratio Chi-Square (53.7054, $p<0.0001$). Although the Mantel-Haenszel Chi-Squared is formulated to assess linear associations among ordinal categories, it also exhibits significance (8.9417, $p=0.0028$). These assessments indicate that the probability of *Wage_Gap_Indicator*=1 (as opposed to 0) systematically varies with a subject's level of formal education. Nevertheless, the effect size metrics (Phi=0.0462, Contingency Coefficient=0.0461, Cramer's V=0.0462) suggest a minimal association where, despite the substantial statistical significance, the strength of the relationship remains modest, indicating that although education level and wage gap status are not independent, they are not highly correlated. This phenomenon, as characterised by exceedingly small effect sizes coinciding with highly significant p-values, transpires when extensive samples render even negligible disparities statistically significant. In conclusion, while the findings refute the hypothesis of independence, the practical impact of general education on wage gap status appears relatively minor within this dataset.

To sum up, an analysis of contingency tables, alongside visualising observed trends over time, assists population researchers in discerning statistically significant variables suitable for incorporation into the longitudinal regression model frameworks, where such selection should be undertaken considering the overarching research question and the various objectives of the research endeavour. For example, the analysis of the *Wage_Gap_Indicator* by gender does not yield evidence of a statistically significant relationship between gender and the wage gap; thus, the gender variable may not qualify as an explanatory variable within a regression model framework. Nevertheless, the specific aims of the research project may encompass an examination of dimensions such as educational attainment and specific occupational categories, as well as the potential existence of a wage gap that is presumed to affect the gender demographics. Consequently, integrating gender into the regression model framework becomes imperative.

The decision regarding variable inclusion in the final regression model is predominantly contingent upon the research objectives rather than exclusively informed by statistical associations among the variables derived from elementary contingency tables; however, analysing contingency tables remains crucial as it furnishes researchers with significant insights necessary for formulating the longitudinal regression model frameworks.

This section focused on the contingency tables corresponding to the various remaining categorical variables within the dataset associated with Case Study 1, including, but not limited to, the dimensions of *General Education, Vocational Qualification, Occupation, Firm Size, Sector, Region, Union Membership, Parental Status,* and *Training Participation.* For conciseness, a comprehensive examination of every contingency table has been excluded. It is recommended that readers undertake these analyses utilising the dataset provided.

The purpose of using PCA as part of an EDA process is to reduce the dimensionality of large datasets while retaining the most significant variables, thereby facilitating efficient data processing and interpretation. PCA achieves this by transforming the original set of continuous variables into a new set of uncorrelated variables, known as principal components, which are linear combinations of the original variables ordered by the amount of variance they capture from the data. This transformation retains most of the variance present in the original dataset, making PCA a valuable tool in EDA and regression. In this section, let us examine the continuous variables encompassed within the dataset of Case Study 1, which include the variables related to wage, tenure, and age, as they hold significant importance for our analytical pursuits. The primary aim we seek to achieve through the implementation of PCA is to explore the potential avenues for reducing the dimensionality of the data encapsulated within these variables, thereby enhancing our understanding of their underlying patterns. Ultimately, this analytical undertaking transcends a technical procedure; it elucidates the complex dynamics associated with the data, thereby promoting a deeper understanding of its structure and enabling more nuanced interpretations.

The SAS FACTOR procedure performs various factor and component analyses and rotations. Input to this procedure can be multivariate data, a correlation matrix, a covariance matrix, a factor pattern, or a scoring coefficient matrix. The employed methodology can decompose the correlation or covariance matrix, and most of the findings can be preserved in the output dataset. A general SAS code for the PCA analysis is provided below.

SAS Syntax 4.11

```
proc factor data=Case_Study_1
        method=principal
        nfact=3
```

```
        out=pca_scores
        scree;
    var Age Job_Tenure Wage;
run;
```

What the Code Does

This syntax runs a PCA analysis on continuous variables of *Age*, *Job Tenure*, and *Wage*, extracting up to three components and saving the component scores to a new dataset. The SCREE option prints the eigenvalue scree plot so you can visually assess how much variance each component captures. In our context, PCA is demonstrative, showing SAS syntax and diagnostic output, because three variables offer little scope for dimension reduction (Tables 4.9–4.12).

TABLE 4.9

The Factor Procedure

Eigenvalues of the Correlation Matrix: Total=3 Average=1				
	Eigenvalue	Difference	Proportion	Cumulative
1	1.14147761	0.14955035	0.3805	0.3805
2	0.99192726	0.12533213	0.3306	0.7111
3	0.86659513		0.2889	1.0000

TABLE 4.10

The Factor Patterns

Factor Patterns			
	Factor1	Factor2	Factor3
Age	0.56460	−0.66953	0.48265
Job_Tenure	0.74429	−0.00565	−0.66784
Wage	0.51841	0.73731	0.43317

TABLE 4.11

The Factor Patterns, Variance

Variance Explained by Each Factor		
Factor1	Factor2	Factor3
1.1414776	0.9919273	0.8665951

TABLE 4.12

The Factor Patterns, Communality

Final Communality Estimates: Total=3		
Age	Job_Tenure	Wage
1.0000000	1.0000000	1.0000000

The results generated by the PROC FACTOR elucidate the outcomes of the PCA factor analysis employing three observed variables (*Age, Job_Tenure, and Wage*) within the Case Study 1 dataset comprising 25,000 records. Given the presence of three variables, the analysis extracts three factors by the NFACTOR criterion, implying the absence of actual dimension reduction (as each factor collectively accounts for all the variance). The table of eigenvalues describes the extent of variance captured by each factor in the original correlation matrix. The foremost aspect illustrates around 38 per cent of the combined variance (eigenvalue ~1.11), the following factor holds approximately 33 per cent (eigenvalue ~0.99), and the last one encompasses nearly 29 per cent (eigenvalue ~0.86). The factor pattern matrix elucidates the strength of correlation of each variable with the three extracted factors. For example, *Age* registers a loading of 0.5646 on Factorl, 0.6695 on Factor2, and 0.4826 on Factor3, indicating moderate positive links with Factorl and Factor3, whilst revealing a negative association with Factor2. *Job_Tenure* exhibits the most substantial loading on Factorl (0.7443) and Factor3 (–0.6876), whereas the *Wage* variable registers high on Factor2 (0.7373) but also shows moderate positive loadings on Factorl and Factor3. The interpretation of these loadings within a factor analysis framework is constrained, as all three factors are retained. Importantly, the final communality estimates for the three variables are 1, signifying that the solution recaptures all of the variance associated with *Age, Job_Tenure,* and *Wage* by retaining as many factors as there are variables. Consequently, the standardised scoring coefficients presented in Table 4.13 illustrate how each variable contributes to calculating each factor score. The output encapsulates the fundamental elements of principal components: factors derived from the data's covariance structure, their corresponding eigenvalues, and the loadings of each variable that demonstrate how the factors correspond to the original variables.

Given these results, including all three variables in an envisaged longitudinal regression model is advisable, as they may each provide unique insights into the outcome variable. However, diagnostics such as the Variance Inflation Factor (VIF) should be conducted to assess multicollinearity among the predictors. If multicollinearity is present, variables like *Age*, which have a lower communality, could be considered for exclusion.

TABLE 4.13

The Factor Patterns, Standardised Scoring Coefficients

	Factor Patterns		
	Factorl	**Factor2**	**Factor3**
Age	0.49462	–0.67498	0.55695
Job_Tenure	0.65204	–0.00570	–0.77065
Wage	0.45415	0.74331	0.49985

Additionally, stepwise regression or adjusted R-Squared, AIC/BIC criteria can help evaluate whether including all variables improves the model's explanatory power. The final decision should always balance statistical results and theoretical justification, ensuring that only variables with meaningful contributions are retained in an envisaged longitudinal regression model.

This chapter examined EDA methodologies with an emphasis on longitudinal datasets. It highlighted the significance of visualising raw data to identify patterns of scientific pertinence relevant to longitudinal investigations and detect anomalous observations. Line plots were emphasised as critical instruments for illustrating responses across temporal dimensions, consequently aiding in identifying trends and variability among subjects. A significant endeavour is allocated to applying descriptive statistics and visual representations in the context of data analysis. This chapter further described discernible patterns and inter-subject variability by establishing associations among repeated measurements for each subject over time, where visualising observed profiles over time was imperative for recognising overarching trends, uncovering non-linear transformations, and understanding variability at specific temporal intervals. Furthermore, the application of individual and collective profiles to promote a deeper understanding of the fundamental patterns and interrelations within the dataset is provided, and it is shown that PCA is such a technique that reduces the dimensionality of extensive datasets while retaining essential features, thus facilitating efficient data processing and interpretation.

Reflection Questions and Answers

Question 1 – You discover that 18 per cent of workers have missing wage observations in one or more years, but only 2 per cent have missing values for *Job Tenure*. How might this asymmetric missingness bias describe wage statistics, and what imputation or weighting strategies could you use to minimise that bias?

Answer 1 – If 18 per cent of wage observations are missing while *Job Tenure* is nearly complete, descriptive wage statistics (means, medians, variance) will likely be biased downward if missingness is higher among low-paid workers, or upward if high earners are more reluctant to report. Two widely used remedies are multiple imputation, drawing plausible wage values from a prediction model that conditions on fully observed covariates such as tenure, age, and sector and inverse-probability weighting, where each complete case is weighted by the inverse of its estimated response probability so that complete cases represent similar incomplete ones. Diagnostics after either method help ensure the adjustment does not distort the underlying variance structure.

Question 2 – A missing-value heat-map shows almost no *Wage* data for 2013 in the public-sector subsample. List two substantive and two data-processing explanations for this pattern, and outline diagnostic checks to distinguish between them.
Answer 2 – Substantive explanations: (a) a wage-freeze or strike settlement delayed 2013 payments, leaving wages recorded at zero; (b) a mid-year payroll-system change caused genuine under-reporting for public employees. Data-processing explanations: (c) a file from the public-sector payroll system was never loaded; (d) the wage field was mislabelled during import and dropped. Diagnostic checks include (i) inspecting the raw 2013 micro-file headers for missing wage columns, (ii) comparing auxiliary earnings variables to see if the gap is specific to the Wage field, (iii) examining audit logs for 2012–2014 file loads, and (iv) validating against an external aggregate to see whether public-sector wages were paid.

Question 3 – When computing within- and between-subject standard deviations, *Wage* shows a between/within ratio of 4.8, while *Age* shows 0.3. What does each ratio imply about the utility of these variables for fixed-effects modelling?
Answer 3 – A ratio of 4.8 signals that most variability in *Wage* is between workers; within-worker wage movements exist but are modest relative to cross-sectional differences. The ratio of 0.3 for *Age* indicates the opposite: nearly all variance is within-subject (people inevitably age), so *Age* offers little cross-sectional discriminatory power. For fixed-effects models, which exploit within-subject change, *Age* will be helpful to (strong within variance). In contrast, *Wage* as a dependent variable may yield a small within R-Squared because limited intra-subject wage fluctuation is available to explain.

Question 4 – The panel's duplicate-record audit (PROC SORT … DUPOUT=) flags 0.7 per cent of person-year rows as duplicates. Discuss the potential consequences of keeping versus removing these duplicates for (a) descriptive histograms and (b) subsequent panel regressions.
Answer 4 – (a) Descriptive histograms: keeping duplicates inflates the weight of specific values, skewing the histogram and potentially exaggerating modes (e.g., many duplicated entries at a round wage figure). (b) Panel regressions: duplicates violate the assumption of independent errors, biasing standard errors downward and potentially doubling the influence of some individuals, which can distort coefficients. Best practice is to remove exact duplicates while retaining genuinely repeated measures, documenting the decision to preserve reproducibility.

Question 5 – Examine the wage histogram with kernel density overlay. If the distribution shows heavy right skew and a long left shelf, what visual or numeric evidence would convince you to log-transform Wage before further analysis?

Answer 5 – Visual cues: a long right tail with sparse but extremely high values; a left shelf of low values hugging the x-axis. Numeric cues: mean > median, skewness > 1, and a considerable distance between the 75th percentile and the median relative to the interquartile range. If a log transform pulls the mean-median gap close and reduces skewness below 0.5, it is justified; otherwise, trimming outliers may suffice.

Question 6 – A PROC FREQ cross-tab of *Wage Gap Indicator* × *Gender* returns a Pearson Chi-Squared *p*-value of 0.48. Describe three reasons why a non-significant bivariate association might warrant multivariate investigation.
Answer 6 – (i) Confounding: gender differences may only emerge once occupation or education is controlled. (ii) Power: The Chi-Squared test may lack sensitivity if event counts are small. (iii) Non-linear effects: the relationship may exist only in specific age bands or sectors. Thus, multivariate models can still uncover structured gender gaps hidden in the bivariate tabulation.

Question 7 – In the occupation-specific wage-trend plot, one occupation shows a sharp wage drop in 2012, followed by a rapid rebound. Name two graphical or statistical checks that could help decide whether this is an actual economic event or a data artefact.
Answer 7 – Graphical check: overlay employment counts; if counts plummet simultaneously, the dip may be data loss. Statistical check: compare median wages in administrative earnings data from an external source; if both series fall, the dip is likely real. Also, test whether the dip appears only in one occupation code version or across all definitions.

Question 8 – After running PCA on *Age, Job Tenure,* and *Wage,* the first component's eigenvalue is 1.95 and its loadings are roughly 0.02 for *Age,* 0.71 for *Job Tenure,* and 0.70 for *Wage.* Interpret this component substantively, and argue for or against retaining it in subsequent analysis.
Answer 8 – Loadings of ~ 0.71 for *Job Tenure* and *Wage* and near-zero for *Age* suggest a tenure-wage composite capturing the joint movement of experience and pay. With an eigenvalue below the Kaiser criterion, the component explains moderate variance. Retain it only if a single synthetic index of tenure-adjusted wage is analytically useful; otherwise, the raw variables convey precise meaning.

Question 9 – Suppose the completeness diagnostic finds that 40 per cent of *Promotion* values are missing, concentrated among part-time workers. Propose a principled approach for handling this pattern, other than simple deletion, and describe how you would document the decision.
Answer 9 – One principled approach is multiple imputation with a missing-at-random assumption, conditional on part-time status and other observables; include interaction terms (*Part-Time* × *Gender,* etc.) in the imputation model. Alternatively, create an explicit "Unknown Promotion" category to

preserve sample size while flagging uncertainty. Document the mechanism (attrition bias among part-timers) and the chosen remedy in a data-issues log.

Question 10 – The timeliness audit shows that the median *EntryDelay* (data-entry lag) shrinks from 90 days in 2010 to 30 days in 2014, but the maximum lag remains above two years. Discuss how long-tail delays could distort year-on-year wage trends and suggest one visual and one numeric diagnostic to assess the impact.

Answer 10 – Long lags can understate recent wages if late entries are systematically high (e.g., bonuses) or overstate them if low earners have more prolonged delays. Visual diagnostic: a box plot of *EntryDelay* by Year showing outliers in recent years. Numeric diagnostic: recompute mean *Wage* for each year, including only records with *EntryDelay* < 180 days; compare the revised means to the full-sample means to quantify drift introduced by late entries.

5

Within-Subject Inference: Fixed-Effects Models for Wage Inequality

To orient readers and maintain the navigational rhythm established in earlier chapters, this chapter begins with explicit sign-posting of its learning objectives. Building directly on the conceptual trade-offs among fixed-effects (FE) and random-effects (RE) models introduced in Chapter 1 and on the wage-panel dataset described in Chapter 3, this chapter will show readers why the FE estimator is preferred when unobserved, time-invariant subject-level attributes (e.g., innate ability) are correlated with observed covariates, how to transform a long-format panel into its within-subject form, and how to fit and interpret linear and conditional-logit FE models using SAS Studio. Along the way, readers will learn to run core diagnostics, quantify the variance share removed by data transformation, assess the impact of dropped units in a conditional-logit context, and, finally, translate statistically significant within-subject effects into understandable labour-market policy insights, all while adhering to the reproducible coding and data-governance principles established in Chapter 2.

Chapter 1 demonstrated that the central methodological tension in longitudinal regression is how we handle unobserved, time-invariant heterogeneity, for example, the subject-specific fixed traits that may or may not correlate with observed traits. The FE estimator is the workhorse that delivers consistent, unbiased coefficients whenever those latent traits are correlated with the regressors, whereas an RE estimator gains efficiency under the assumption of zero correlation.

Our wage panel of the Case Study 1 fits the FE remit from this perspective, where innate ability, motivation, and firm-matching quality, none observed in the administrative data, are correlated with observed variables such as education, tenure, and union status. Hence, if we treat those unobserved traits as orthogonal RE, the RE estimator would attribute part of their influence to the covariates, biasing policy-relevant parameters (e.g., the return to vocational training). By stripping out the subject mean, FE shields population researchers from that bias and enables the interpretation of β as the within-subject effect, for example, the change in wage associated with a one-unit change in the covariate, while holding all time-invariant personal features constant.

The concept of FE regressions is to utilise each subject as their control, thus concentrating on within-subject variability to derive the regression coefficients. By assessing each subject's data at various time points across the

 DOI: 10.1201/9781003599869-5

longitudinal continuum, FE models account for stable traits of that subject, regardless of whether such traits are observed or not. This is because any characteristic that does not change over time will influence all measurements uniformly; hence, its impact is removed when determining the change over time for each subject. Simply put, FE models discard the influence of variables that remain constant over time by viewing them as fixed parameters that do not require direct estimation. This is accomplished by emphasising changes within subjects rather than the differences between them. Consequently, any trait that remains unchanged over time does not factor into the analysed variation, effectively mitigating its impact.

However, some limitations are associated with this approach, such as estimating large standard errors since they rely solely on within-subject variations while disregarding information about between-subject differences that lead to increased *p*-values and wider confidence intervals, especially when predictors show variability across subjects but exhibit little change over time for each subject. Although FE models result in greater standard errors, they may yield accurate evaluations of predictors' effects that change over time as they isolate the variation that can be directly attributed to changes within subjects, which is often pertinent for grasping causal relationships.

The following presents the mathematical representation for a dependent variable y_{it}, which varies over time and across subjects:

$$y_{it} = \alpha_i + \beta x_{it} + \varepsilon_{it}$$

where

- β: Vector of coefficients for the predictor variables.
- α_i: The fixed effect, capturing subject-specific characteristics that are constant over time.
- x_{it}: The vector of independent variables for subject *i* at time *t*.
- ε_{it}: The error term, capturing time-varying unobserved factors.

A central underlying assumption of FE models asserts that the characteristics of subjects that remain unchanged over time are effectively controlled within the analytical structure. This is achieved by treating these characteristics as fixed parameters, either estimated directly or systematically excluded from the estimation process. This assumption enables the model to focus on how variables change over time, effectively showcasing the influence of the most important variable. This goal is reached through intra-subject comparisons, which require at least two measurements from each subject. The variations in these measurements are then averaged across the entire sample population. This methodological strategy aids in reducing the effect of stable subject characteristics, ensuring that the observed effects can be attributed to changes over time rather than inherent differences between subjects. Second, in FE models' error terms, which represent random variability, are presumed to have an expected value of zero. This assumption is consistent with other linear models, wherein error terms are treated as random variables defined by a probability distribution, often assumed to be normal with a mean of zero and a specified variance. The assumption of no perfect multicollinearity among the independent variables of the FE model constructs allows for the estimation of each variable's effect, whilst the assumption of homoscedasticity posits that the variance of the error term remains consistent across all observations ensuring that the estimates produced by the models are efficient and unbiased. Needlessly to say that when these assumptions are violated, skewed estimates and faulty conclusions emerge.

Implementation-wise, FE regressions can be conducted using various SAS procedures, including PROC GLM for linear models, PROC LOGISTIC for logistic regression models, PROC GENMOD for count data models, and PROC PHREG for survival models. These procedures enable researchers to employ FE models across chosen statistical frameworks.

As mentioned earlier, FE estimations serve as the foundation for causal analysis when unobserved, time-invariant characteristics pose a risk of confounding a linear regression. By adjusting the study design so that each variable acts as its control, FE models then effectively eliminate latent attributes such as genetic predisposition, initial motivation, and organisational culture that remain constant within a subject (or firm, school, or region). Multiple algebraically equivalent quantitative methods yield the exact FE model estimates for a numeric outcome, such as (i) entity demeaning involves subtracting each subject's average from every observation, (ii) least-squares dummy variable regression incorporates an intercept for each subject, and (iii) first-differencing is applicable when the longitudinal panel consists of two waves. These methods can be estimated using ordinary least squares (OLS) and their data requirements depend on: (i) a minimum of two observations per subject on the outcome, and (ii) some within-subject variation in the variables. A design that meets these criteria is a before and after study, for example, measuring subjects' weight 1 week before and 1 week after following a diet programme, to isolate the intervention's effect while accounting for each subject's inherent metabolism. In a two-wave scenario, this construct simplifies to a paired t-test, and in panels with additional waves, the same hypothesis is evaluated using the conventional t-statistics and F-statistics, ideally with cluster-robust standard errors to adjust for serial correlation in the residuals. These tests determine whether the observed intra-subject changes, such as the average kilograms lost post-diet or the wage increase following on-the-job training, are statistically significant to exclude chance as a likely explanation.

As outlined in Chapter 4, before implementing any FE model, population researchers need to initiate a diagnostic review of the datasets that begins by outlining the panel's structure: determine the number of distinct individuals, count the available waves, and calculate the average observations per subject. Subsequently, a missing data dashboard for every outcome and covariate should be developed, as even a small percentage of attrition can skew within-subject estimates. Lastly, population researchers compile the within versus between-subject standard deviations for each predictor and outcome. These preliminary checks affirm that the data can support a credible FE specification.

The remainder of this section illustrates an FE model development workflow using the data from Case Study 1, where the continuous variable wage acts as the dependent measure of labour-market reward. By implementing the most straightforward within transformation to this long-format panel, we demonstrate how FE estimations can differentiate the intra-subject wage response to variations in selected subject characteristics.

As noted earlier, Case Study 1 endeavours to elucidate the methodology for implementing an FE model utilising SAS Studio to examine how socio-economic factors contribute to wage disparities over a temporal continuum. Hence, the dependent variable in our envisaged FE model construct is the wage, a continuous variable that captures the annual gross wage in EUR. The vector of selected independent variables for the subjects *Age, Firm_Size, General_Education,* and *Occupation.* The SAS Syntax 5.1 generates the output for this analysis.

SAS Syntax 5.1

```
proc sort data=Case_Study_1 out=panel_sorted;
  by ID; run;
proc glm data=panel_sorted;
  absorb ID;               /* removes subject-specific intercepts */
  class Year Firm_Size General_Education Occupation;
  model Wage =
     Age
     Firm_Size General_Education Occupation Year
     / solution;
quit;
```

What the Code Does

In this PROC GLM call, the ABSORB ID option instructs SAS to remove each subject-specific intercept, effectively applying the within transformation without explicitly creating thousands of dummy variables. The accompanying CLASS statement designates the categorical predictors as reference factors; hence, PROC GLM can generate the necessary dummy variables behind the scenes. The SOLUTION option requests the coefficient estimates, their standard errors, and *t*-statistics, giving readers the statistical output for every within-subject effect inherent.

By absorbing each person's unconditional mean, the model isolates within-subject wage movements as these covariates change, thereby purging the influence of unobserved, time-invariant attributes such as innate ability or family background. SAS PROC GLM statements for estimating the basic model include a CLASS statement to handle the categorical variables. Note that for this to work, the data set must be sorted by the variable specified on the ABSORB statement. The SOLUTION option tells GLM to print the coefficient estimates and their associated statistics. The resulting estimates (reported in Tables 5.1–5.5) quantify the average change in a subject's wage when they age 1 year, switch to a different firm-size category, obtain an educational attainment, or change occupation, while holding constant all other characteristics that remain fixed for that subject.

Table 5.1 demonstrates the model fit statistics. The comparison of the Model Sum of Squares (MSS) to the Error Sum of Squares (ESS) yields an F-value of 6.45 ($p<0.0001$), indicating that the model demonstrates a high

TABLE 5.1

Model Fit Statistics I

Source	DF	Type I SS	Mean Square	F-Value	Pr>F
Model	5019	940746201374	187436979.75	6.45	<.0001
Error	19980	580984023888	29078279.474		
Corrected Total	24999	1.5217302E12			

TABLE 5.2

Model Fit Statistics II

R-Square	Coeff Var	Root MSE	Wage Mean
0.618208	13.95598	5392.428	38638.85

TABLE 5.3

Type I Sums of Squares

Source	DF	Type I SS	Mean Square	F-Value	Pr>F
ID	4999	927356225202	185508346.71	6.38	<.0001
Age	1	12989099187	12989099187	446.69	<.0001
Firm_Size	2	14090024.415	7045012.2075	0.24	0.7848
General_ Education	5	129525503.51	25905100.702	0.89	0.4860
Occupation	9	219846474.02	24427386.002	0.84	0.5790
Year	3	37414982.567	12471660.856	0.43	0.7323

TABLE 5.4

Type III Sums of Squares

Source	DF	Type II SS	Mean Square	F-Value	Pr>F
Age	0	0.0			
Firm_Size	2	13922065.6	6961032.8	0.24	0.7871
General_ Education	5	132762183.2	26552436.6	0.91	0.4712
Occupation	9	219207707.0	24356411.9	0.84	0.5812
Year	3	37414982.6	12471660.9	0.43	0.7323

degree of statistical significance. The R-square value of 0.618208 indicates that approximately 61.82 per cent of the variance in Wage is accounted for by the predictors included, which is considerable, albeit some variance remains unexplained. The root mean square error (RMSE) of 5392.428 denotes the square root of the residual variance, where a lower RMSE suggests a compact clustering of residuals around the fitted regression line. The RMSE value needs to be interpreted about the scale of the wage variable.

Type I sums of squares (SS), presented in Table 5.2 for the variables *ID, Age, Firm_Size, General_Education, Occupation,* and *Year,* show the sequential contribution of each factor to the variation in wages. Notably, ID exhibits a substantial sum of squares, indicating that variability at the individual level constitutes a significant proportion of wage discrepancies. In contrast, Type III SS, shown in Table 5.4, demonstrates the distinct contribution of each factor while simultaneously controlling for the influence of all other variables. As is noted, *Age* is the most significant predictor, maintaining its statistical significance even after adjusting for *Firm_Size, General_Education, Occupation,* and *Year* whilst *Firm_Size* does not indicate statistical significance, suggesting that the size classification of the firm (e.g., small, medium, or large) does not independently affect wage disparities when considering other influences. In the same manner, neither *General_Education* nor *Occupation* reaches a level of statistical significance, which means these variables cannot be deemed effective wage predictors after factoring in *Age* and other variables. Similarly, *Year* exhibits no statistically significant effect when other variables have been controlled for.

In examining the parameter estimates of Table 5.5, the variable *Age* is associated with a coefficient of 543.1977 ($p<0.0001$), suggesting that each increment of 1 year in age correlates with an average wage increase of approximately EUR543. The *t*-value of 5.04 reinforces the notion of *Age* as a robust positive predictor of wage. With respect to the *Firm_Size,* the disparities between large and small enterprises (79.83, $p=0.5886$) and between medium and small enterprises (–3.57, $p=0.9735$) lack statistical significance, indicating an absence of a meaningful wage premium. *General_Education* levels, similarly, demonstrate lack of wage differentials when other variables are controlled for.

Table 5.5

Parameter Estimates

Parameter	Estimate	Standard Error	t-Statistics	Pr>|t|
Age	543.1977390	107.8820642	5.04	<.0001
Firm_Size Large	79.8340664	147.6143877	0.54	0.5886
Firm_Size Medium	–3.5674263	107.3225125	–0.03	0.9735
Firm_Size Small	0.0000000			
General_Education Certificate of Ten-Grade School	563.8019474	405.4578161	1.39	0.1644
General_Education Intermediate School Certificate	73.1877131	243.8452336	0.30	0.7641
General_Education No Secondary School Certificate	607.4503436	478.1759452	1.27	0.2040
General_Education Presently Attending Schools	236.9752706	456.1017171	0.52	0.6034
General_Education Secondary School Certificate	338.4130150	227.5283881	1.49	0.1369
General_Education University Entrance Qualification	0.0000000			
Occupation Armed Forces Occupations	544.6744628	647.2231516	0.84	0.4000
Occupation Clerical Support Workers	424.7992474	483.4892041	0.88	0.3796
Occupation Craft-Related Trades Workers	935.4434783	596.0622569	1.57	0.1166
Occupation Elementary Occupations	522.7995944	348.5478805	1.50	0.1336
Occupation Managers	180.0644615	409.4542355	0.44	0.6601
Occupation Plant and Machine Operators, and Assemblers	82.6824758	344.6869819	0.24	0.8104
Occupation Professionals	460.8362288	305.7792729	1.51	0.1318
Occupation Service and Sales Workers	–184.0388255	443.5144555	–0.41	0.6782
Occupation Skilled Agricultural, Forestry, and Fishery Workers	218.5771022	470.0058549	0.47	0.6419
Occupation Technicians and Associate Professionals	0.0000000			
Year 2010	93.2726268	388.9566373	0.24	0.8105
Year 2011	142.8606272	285.4369528	0.50	0.6167
Year 2012	–3.9124126	186.8752054	–0.02	0.9833
Year 2013	0.0000000			
Year 2014	0.0000000			

Moreover, while a theoretical rationale for disparities in occupational wages may exist, the data at hand do not reveal substantial variations once other controls are applied. Likewise, the *Year* dummy variables yield exceedingly low coefficients alongside elevated *p*-values, signifying the absence of a discernible temporal effect on wages.

The within-subject estimates identify *Age* as the dominant source of wage variation: every additional year is associated with roughly a 2-per-cent earnings premium, while the unique contributions of *Firm Size, General Education, Occupation,* and calendar *Year* shrink to statistical triviality once *Age* is in the equation. However, before drawing policy conclusions, population researchers should probe whether omitted variables, such as sector, region, detailed job tenure, collective-agreement coverage, or performance bonuses, can explain the residual dispersion prevalent in model diagnostics. Iteratively refining the specification may reallocate explanatory power away from *Age* and yield a model that is more useful for targeted wage-setting policy strategies.

One of the advantages of FE modelling frameworks is its ability to enable population researchers to study whether the age–wage gradient shifts across business cycles or policy regimes. An avenue such as study can achieve this by interacting *Age* with *Year,* where such interaction will capture any calendar-specific changes. SAS Syntax 5.2 shows how readers can set up this construct in SAS Studio.

SAS Syntax 5.2

```
proc glm data=panel_sorted;
  absorb ID;
  class Year Firm_Size General_Education Occupation;
  model Wage =
      Age Firm_Size General_Education Occupation Year
      Age*Year                    /* time-varying age slope */
      / solution;
quit;
```

What the Code Does

The ABSORB ID option removes every subject-specific intercept, turning the model into a within-subject FE regression without the overhead of thousands of dummy variables. Next, the CLASS statement instructs SAS Studio to treat *Year, Firm Size, General Education,* and *Occupation* as categorical factors and automatically create the requisite reference-coded dummies. The explicit *Age*Year* term assigns a separate age slope to each calendar year, so the regression can reveal whether the experience premium drifts over time, whilst the joint F-test on these interaction coefficients indicates whether any such drift is statistically significant. Finally, appending the SOLUTION option instructs PROC GLM to display the full suite of coefficient estimates, standard errors, and t-statistics.

Tables 5.6–5.10 show the results of SAS Syntax 5.2.

TABLE 5.6

Model Fit Statistics I

Source	DF	Sum of Squares	Mean Square	F-Value	Pr>F
Model	5023	940862876863	187310945.03	6.44	<.0001
Error	19976	580867348399	29078261.334		
Corrected Total	24999	1.5217302E12			

TABLE 5.7

Model Fit Statistics II

R-Square	Coeff Var	Root MSE	Wage Mean
0.618285	13.95597	5392.426	38638.85

TABLE 5.8

Type I Sums of Squares

Source	DF	Type I SS	Mean Square	F-Value	Pr>F
ID	4999	927356225202	185508346.71	6.38	<.0001
Age	1	12989099187	12989099187	446.69	<.0001
Firm_Size	2	14090024.415	7045012.2075	0.24	0.7848
General_ Education	5	129525503.51	25905100.702	0.89	0.4860
Occupation	9	219846474.02	24427386.002	0.84	0.5790
Year	3	37414982.567	12471660.856	0.43	0.7323
Age*Year	4	116675489.3	29168872.324	1.00	0.4043

TABLE 5.9

Type II Sums of Squares

Source	DF	Type III SS	Mean Square	F-Value	Pr>F
Age	0	0.0			
Firm_Size	2	13242579.1	6621289.6	0.23	0.7964
General_ Education	5	133573223.6	26714644.7	0.92	0.4675
Occupation	9	216728230.9	24080914.5	0.83	0.5900
Year	3	17471541.0	5823847.0	0.20	0.8962
Age*Year	4	116675489.3	29168872.3	1.00	0.4043

TABLE 5.10

Parameter Estimate

Parameter	Estimate	Standard Error	*t*-Statistics	Pr>|*t*|
Age	666.4613388	361.548186	1.84	0.0653
Firm_Size Large	76.7209490	147.642141	0.52	0.6033
Firm_Size Medium	−4.8016365	107.342973	−0.04	0.9643
Firm_Size Small	0.0000000			
General_ Education Certificate of Ten-Grade School	565.9742799	405.466236	1.40	0.1628
General_ Education Intermediate School Certificate	70.8541353	243.870824	0.29	0.7714
General_ Education No Secondary School Certificate	612.6527186	478.195148	1.28	0.2001
General_ Education Presently Attending Schools	237.0959373	456.117081	0.52	0.6032
General_ Education Secondary School Certificate	336.6683947	227.561063	1.48	0.1390
General_ Education University Entrance Qualification	0.0000000			
Occupation Armed Forces Occupations	543.0892664	647.298837	0.84	0.4015
Occupation Clerical Support Workers	420.7064554	483.500768	0.87	0.3842
Occupation Craft-Related Trades Workers	929.1349863	596.086684	1.56	0.1191
Occupation Elementary Occupations	522.7802441	348.560755	1.50	0.1337
Occupation Managers	181.4367020	409.498440	0.44	0.6577

(Continued)

TABLE 5.10 (*Continued*)

Parameter Estimate

Parameter	Estimate	Standard Error	*t*-Statistics	Pr>|*t*|
Occupation Plant and Machine Operators, and Assemblers	89.0988314	344.710351	0.26	0.7960
Occupation Professionals	462.9592442	305.800397	1.51	0.1301
Occupation Service and Sales Workers	−180.8322260	443.561326	−0.41	0.6835
Occupation Skilled Agricultural, Forestry, and Fishery Workers	217.2529587	470.021141	0.46	0.6439
Occupation Technicians and Associate Professionals	0.0000000			
Year 2010	55.0913461	1309.866403	0.04	0.9665
Year 2011	181.6448460	959.490607	0.19	0.8498
Year 2012	230.5395842	626.189120	0.37	0.7128
Year 2013	0.0000000			
Year 2014	0.0000000			
Age*Year 2010	12.8739661	7.797424	1.65	0.0987
Age*Year 2011	7.8326274	7.796931	1.00	0.3151
Age*Year 2012	0.2781198	7.798417	0.04	0.9716
Age*Year 2013	2.7834622	7.796675	0.36	0.7211
Age*Year 2014	0.0000000			

A primary emphasis in this analysis is on the *Age* × *Year* interaction, which is incorporated as a separate set of predictors, with one parameter allocated for each year after the reference year. The Type III SS table reveals an F-value of 1 for the collective examination of all *Age* × *Year* interaction parameters. This finding suggests that there is no statistically significant evidence to support the notion that the influence of Age on wages varies across the years. The parameter estimates table, on the other hand, lists the estimates and *p*-values associated with each *Age* × *Year* term (e.g., *Age* × *Year 2010*, *Age* × *Year 2011*), where each estimate represents the incremental variation in the slope of Age for the respective year compared to the reference year category (which is the Year 2014). As is seen, all *Age* × *Year* estimates exhibit large standard errors and yield *p*-values surpassing conventional significance thresholds. These findings suggest that the relationship between age and wages remains stable mainly over the years.

Upon introducing interaction terms with *Year*, as noted in Table 5.10, the coefficient for *Age* registers at 666.4133 with a *p*-value of 0.0653. While still indicative of a positive relationship between *Age* and wages, its significance is marginally diminished compared to baseline models. This change arises because some of the variability previously attributed exclusively to *Age* is now partitioned into the *Age* × *Year* interaction parameters. This suggests that each additional year of age is associated with an average wage increase of EUR666, albeit with disappearing statistical significance when year-by-year changes are considered.

Our FE analysis shows that the contrasts between *Firm-Size* exhibit t-statistics that indicate that it does not provide any variation in wages among subjects once age is accounted for. The same trend is observed in the contrasts related to *General-Education* as well as across the broader *Occupation* groups, where the *p*-values consistently exceed conventional thresholds, implying that transitions between these categories are either infrequent among subjects or lack correlation with wage fluctuations. Further, *Year* dummies prove to be jointly statistically insignificant. In summary, after removing time-invariant heterogeneity, our FE model assigns nearly all explainable wage variation to the ageing/experience dimension.

When a binary outcome is analysed with an FE logistic regression, each subject-specific intercept becomes an additional parameter as the number of intercepts grows with the sample size, the envisioned model would violate the regularity conditions of maximum-likelihood estimation theory (an issue known as the incidental-parameters problem), and absorbing these effects with dummy variables would yield biased slope estimates. Therefore, population researchers would estimate the likelihood conditioned on a sufficient statistic, rather than estimating every intercept, known as conditional (FE) maximum likelihood.

In SAS, conditional (FE) maximum likelihood is implemented with PROC LOGISTIC … STRATA ID statements, which fits a conditional logit to the data. The conditional (FE) maximum-likelihood estimator retains the desirable sample properties of consistency and asymptotic normality, thus offering a statistically rigorous solution to the incidental-parameters problem in longitudinal logit models.

Case Study 1 operationalises wage inequality with a binary *Wage_Gap_Indicator* that is generated, for each calendar year, as we computed the median wage in every *Gender* × *Occupation* × *Education Level*, attaining each subject record with a value of 1 if their wage falls below this peer-group benchmark and 0 otherwise. We then model the within-subject odds of falling below the median as a function of time-varying circumstances, for example, *Vocational Qualification, Sector, Region, Promotion Status*, and a vector of *Year* dummies. Subject-fixed intercepts are eliminated via conditional likelihood, so coefficients can be interpreted as the effect of changing one of these attributes on the probability of a wage gap, holding constant all time-invariant subject traits.

SAS Syntax 5.3

```
/* Conditional FE logistic regression */
proc logistic data=Case_Study_1_panel desc;
  class Year Vocational_Qualification Sector
      Region Promotion/param=ref;
  model Wage_Gap_Indicator(event='1') =
      Year Vocational_Qualification Sector
      Region Promotion;
  strata ID;
```

What the Code Does

The DESC option directs SAS to model the probability that a subject earns below the demographic-peer median. The CLASS statement, paired with / PARAM=REF, transforms every categorical predictor into reference-coded dummies so that each estimated coefficient reflects the log-odds change relative to its baseline category. Finally, STRATA ID activates the conditional ML estimator, which differences out all subject-specific intercepts and automatically drops subjects whose wage-gap status never varies over time, ensuring that only within-subject information contributes to the likelihood.

The resulting coefficients quantify, for example, how moving from non-vocational to vocational qualification or switching from the public to the private sector alters a subject's risk of earning below their demographic-peer median, providing insights that would feed into the equity-oriented policy discussions. Let's now turn our attention to the analysis of SAS Syntax 5.3 output tables.

Table 5.11 shows the data categorised into various response patterns, each accompanied by a corresponding number of strata and frequency. Such a summary is commonly encountered in conditional logistic regression, for example, within matched case-control frameworks or stratified analytical approaches. Each stratum delineates a cohort within which the observations are subjected to

TABLE 5.11

Strata Summary

Response Pattern	Wage_Gap_ Indicator(1)	Wage_Gap_ Indicator(0)	Strata	Frequency
1	0	5	834	4170
2	1	4	924	4620
3	2	3	918	4590
4	3	2	893	4465
5	4	1	789	3945
6	5	0	642	3210

comparative analysis; however, the specifics of the matching or stratification criteria are derived from the overarching study design by the population researchers. The salient conclusion that readers can gain from the analysis of Table 5.11 is that the model has duly incorporated these strata in estimating parameters, thereby mitigating the influence of latent confounding variables.

Omnibus statistics, as shown in Table 5.12, test the null hypothesis that every slope parameter equals zero once subject-fixed effects have been controlled for. With 27 degrees of freedom, the likelihood-ratio Chi-squared of 44.8 (p=0.017) rejects the null hypothesis whilst the Score and Wald tests corroborate the result with p-values of 0.015 and 0.023, respectively. In other words, the block of time-varying covariates of *Year, Vocational Qualification, Sector, Region,* and *Promotion Status,* collectively improves model fit relative to a baseline. The FE logit, therefore, captures at least one substantive determinant of a subject's risk of earning below the peer-group median wage.

Table 5.13 examines the individual coefficient estimates to identify which predictors contribute to this global significance and to translate their log-odds effects into policy-relevant probability changes. Given that our analysis employs a logistic regression model, each coefficient signifies the log-odds alteration in *Wage_Gap_Indicator*=1 (for instance, the existence of a wage gap) resulting from a one-unit variation in the predictor variable (if continuous) or affiliation with a specific category (if categorical), as benchmarked against the baseline. The years 2010, 2011, 2012, and 2013 exhibit minor coefficients and large p-values.

A negative log-odds coefficient estimate indicates that subjects with a Bachelor's degree are characterised by a diminished likelihood of experiencing a wage gap compared to the reference group. Correspondingly, obtaining a Master's degree is associated with a considerably lower chance of experiencing a wage gap. While the Doctor's Degree, and No Vocational Qualification categories yield p-values exceeding conventional significance thresholds, indicating a lack of statistically significant divergence from the reference group. Although the variable Sector did not demonstrate statistical significance, the parameter estimates reveal coefficients with p-values surpassing 0.05, thereby providing no robust evidence for sectoral disparities in wage-gap odds once other variables have been controlled for. Further, none of the regions exhibits substantial differences from the reference region, with coefficients approaching zero coupled with significant p-values. Lastly, the estimate for *Promotion Status* shows no evidence that promotions influence the wage-gap outcome.

TABLE 5.12

Testing Global Null Hypothesis

Test	Chi-Square	DF	Pr>ChiSq
Likelihood Ratio	44.8024	27	0.0171
Score	45.1313	27	0.0157
Wald	43.5017	27	0.0233

TABLE 5.13

Conditional ML Estimates

Parameter		DF	Estimate	Standard Error	Wald	Pr>ChiSq
Year	2010	1	0.0140	0.0476	0.0864	0.7689
Year	2011	1	0.00774	0.0476	0.0265	0.8707
Year	2012	1	0.0197	0.0476	0.1715	0.6788
Year	2013	1	0.0243	0.0476	0.2603	0.6099
Vocational_Qualification	Apprenticeship	1	0.0685	0.1249	0.3005	0.5836
Vocational_Qualification	Bachelor's Degree	1	–0.9956	0.2550	15.2474	<.0001
Vocational_Qualification	Doctor's Degree	1	0.0741	0.3518	0.0443	0.8332
Vocational_Qualification	Master's Degree	1	–0.8225	0.2795	8.6615	0.0033
Vocational_Qualification	No Vocational Qualification	1	0.0638	0.1358	0.2209	0.6384
Sector	Agriculture Sector	1	–0.1513	0.4104	0.1359	0.7124
Sector	Industry Sector	1	0.1746	0.0863	4.0964	0.0430
Region	Baden-Wurttemberg	1	0.0205	0.2697	0.0058	0.9394
Region	Bayern	1	0.0514	0.2631	0.0381	0.8452
Region	Berlin	1	0.1943	0.2990	0.4224	0.5158
Region	Brandenburg	1	0.0613	0.3313	0.0342	0.8532
Region	Bremen	1	0.7545	0.5292	2.0325	0.1540
Region	Hamburg	1	–0.1167	0.3438	0.1152	0.7343
Region	Hessen	1	0.1299	0.2868	0.2051	0.6506
Region	Mecklenburg-Vorpommern	1	–0.1101	0.3758	0.0858	0.7696
Region	Niedersachsen	1	0.0526	0.2741	0.0368	0.8480
Region	Nordrhein-Westfalen	1	0.0309	0.2606	0.0141	0.9055
Region	Rheinland-Pfalz	1	0.1456	0.2973	0.2397	0.6244
Region	Saarland	1	–0.2755	0.4367	0.3980	0.5281
Region	Sachsen	1	–0.0913	0.2941	0.0963	0.7563
Region	Sachsen-Anhalt	1	0.0997	0.3400	0.0860	0.7693
Region	Schleswig-Holstein	1	0.0307	0.3175	0.0093	0.9230
Promotion Status	0	1	0.00811	0.0567	0.0204	0.8864

The odds-ratio projections, as shown in Table 5.14, confirm that calendar year exerts no material influence on below-median earnings. By contrast, educational attainment shows a transparent gradient; subjects holding a Bachelor's degree realise about 37 per cent of the odds of earning below their peer-group median relative to those with trade or technical credentials (e.g., OR=0.3369, CI 0.224–0.609). The remaining qualification categories, including apprenticeships and doctorates, have wide intervals that overlap one, so their effects are not statistically significant. Sectoral affiliation is mainly neutral, with the sole exception of the industry sector, whose odds of a wage gap are about 19 per cent higher than in services (e.g., OR=1.19, CI 1.01–1.41). None of the regional contrasts with the reference state of Thuringia underscores the lack of differentials once education, sector, and fixed subject effects are controlled. Finally, the promotion dummy is inert, suggesting that within-subject promotions do not systematically shift a subject's risk of falling below the demographic-group median wage. Taken together, the results highlight tertiary education as the principal lever for escaping wage-gap status, while temporal, regional, and most sectoral factors play at best a marginal role after accounting for stable subject attributes.

While conditional FE logit eliminates the effects of time-invariant explanatory variables, it also estimates how these attributes influence the effects of time-varying variables. An important aspect in this context is to combine a stable characteristic with a variable that fluctuates over time, thus forming an interaction term that exhibits variation within subjects. With the following SAS syntax, we will examine whether the impact of receiving a promotion varies between employees in the Industry and Services sectors by incorporating the interaction *Sector* × *Promotion*. The primary effect of Sector will be absorbed and is therefore not estimated. Still, the interaction term reflects the additional within-subject change in the likelihood of a wage gap when a promotion takes place in each sector.

SAS Syntax 5.4

```
proc logistic data=Case_Study_1 desc;
   class Year Vocational_Qualification Sector
        Region Promotion / param=ref;
   model Wage_Gap_Indicator(event='1') =
        Year Vocational_Qualification
        Region
        Promotion
        Sector*Promotion
        / solution;
   strata ID;
run;
```

TABLE 5.14
Odds-Ratio Estimates

Effect	Estimate	95% Wald Confidence Limits	
Year 2010 vs 2014	1.014	0.924	1.113
Year 2011 vs 2014	1.008	0.918	1.106
Year 2012 vs 2014	1.020	0.929	1.120
Year 2013 vs 2014	1.025	0.933	1.125
Vocational_Qualification Apprenticeship vs Trade and Technical School	1.071	0.838	1.368
Vocational_Qualification Bachelor's Degree vs Trade and Technical School	0.369	0.224	0.609
Vocational_Qualification Doctor's Degree vs Trade and Technical School	1.077	0.540	2.146
Vocational_Qualification Master's Degree vs Trade and Technical School	0.439	0.254	0.760
Vocational_Qualification No Vocational Qualification vs Trade and Technical School	1.066	0.817	1.391
Sector Agriculture Sector vs Services Sector	0.860	0.385	1.921
Sector Industry Sector vs Services Sector	1.191	1.006	1.410
Region Baden-Wurttemberg vs Thuringen	1.021	0.602	1.732
Region Bayern vs Thuringen	1.053	0.629	1.763
Region Berlin vs Thuringen	1.214	0.676	2.182
Region Brandenburg vs Thuringen	1.063	0.555	2.035
Region Bremen vs Thuringen	2.127	0.754	6.000
Region Hamburg vs Thuringen	0.890	0.454	1.746
Region Hessen vs Thuringen	1.139	0.649	1.998
Region Mecklenburg-Vorpommern vs Thuringen	0.896	0.429	1.871
Region Niedersachsen vs Thuringen	1.054	0.616	1.804
Region Nordrhein-Westfalen vs Thuringen	1.031	0.619	1.719
Region Rheinland-Pfalz vs Thuringen	1.157	0.646	2.072
Region Saarland vs Thuringen	0.759	0.323	1.787
Region Sachsen vs Thuringen	0.913	0.513	1.624
Region Sachsen-Anhalt vs Thuringen	1.105	0.567	2.151
Region Schleswig-Holstein vs Thuringen	1.031	0.553	1.921
Promotion 0 vs 1	1.008	0.902	1.127

TABLE 5.15
Strata Summary

Response Pattern	Wage_Gap_ Indicator(1)	Wage_Gap_ Indicator(0)	Strata	Frequency
1	0	5	834	4170
2	1	4	924	4620
3	2	3	918	4590
4	3	2	893	4465
5	4	1	789	3945
6	5	0	642	3210

TABLE 5.16
Testing Global Null Hypothesis

Test	Chi-Square	DF	Pr>ChiSq
Likelihood Ratio	46.2143	26	0.0086
Score	46.5577	26	0.0079
Wald	44.9261	26	0.0120

Table 5.15 presents the summary of strata across he response patterns with respect to the Wage_Gap_Indicator showing a consistency between strata and frequency counts while Table 5.16 indicates that our model is statistically significant as all tests' p-values are below 0.05 rejecting the global null hypothesis

What the Code Does

The term *Sector*Promotion* introduces an interaction that varies over time. With this interaction in place, the leading Promotion coefficient captures the within-subject wage-gap effect of receiving a promotion in the reference sector, while the interaction coefficient shows how that promotional effect differs in the Industry sector. Apart from adding this interaction, the syntax is identical to the baseline specification: STRATA ID still conditions out the subject-specific intercepts so that only within-subject changes contribute to the likelihood. This specification, therefore, allows us to answer the question of *"Does a promotion confer the same wage-gap reduction in Industry as it does in Services, or is the benefit sector-specific?"*.

The likelihood-ratio, Score and Wald statistics all reject the joint null hypothesis that every slope, including the *Sector × Promotion* interaction, equals zero (p-values < 0.05). Table 5.17 shows that the coefficient block for the Year dummies remains flat, where estimates cluster around zero and their p-values all exceed 0.10, confirming that once a subject's fixed effects and the other covariates are in the model, calendar time adds no explanatory power to wage-gap status.

TABLE 5.17

Conditional ML Estimates

Parameter		Estimate	Standard Error	Wald	Pr>ChiSq
Year	2010	0.0143	0.0476	0.0898	0.7644
Year	2011	0.00724	0.0476	0.0231	0.8792
Year	2012	0.0202	0.0476	0.1803	0.6711
Year	2013	0.0240	0.0476	0.2530	0.6150
Vocational_Qualification	Apprenticeship	0.0664	0.1251	0.2819	0.5955
Vocational_Qualification	Bachelor's Degree	–0.9960	0.2549	15.2672	<.0001
Vocational_Qualification	Doctor's Degree	0.0785	0.3511	0.0500	0.8231
Vocational_Qualification	Master's Degree	–0.8265	0.2795	8.7452	0.0031
Vocational_Qualification	No Vocational Qualification	0.0638	0.1359	0.2201	0.6390
Sector	Agriculture Sector	–0.1457	0.4104	0.1261	0.7225
Sector	Industry Sector	0.1767	0.0863	4.1927	0.0406
Region*Promotion	Baden-Wurttemberg	–0.0436	0.1048	0.1728	0.6776
Region*Promotion	Bayern	–0.0158	0.1013	0.0245	0.8756
Region*Promotion	Berlin	0.1790	0.1619	1.2221	0.2690
Region*Promotion	Brandenburg	–0.0521	0.1927	0.0730	0.7870
Region*Promotion	Bremen	0.3745	0.4011	0.8717	0.3505
Region*Promotion	Hamburg	–0.2180	0.2167	1.0128	0.3142
Region*Promotion	Hessen	–0.0841	0.1373	0.3747	0.5405
Region*Promotion	Mecklenburg-Vorpommern	0.0129	0.2544	0.0026	0.9595
Region*Promotion	Niedersachsen	0.0545	0.1221	0.1989	0.6556
Region*Promotion	Nordrhein-Westfalen	0.0148	0.0861	0.0297	0.8632
Region*Promotion	Rheinland-Pfalz	0.2496	0.1533	2.6492	0.1036
Region*Promotion	Saarland	–0.0633	0.2923	0.0468	0.8286
Region*Promotion	Sachsen	0.00424	0.1572	0.0007	0.9785
Region*Promotion	Sachsen-Anhalt	–0.0550	0.2069	0.0705	0.7905
Region*Promotion	Schleswig-Holstein	0.1134	0.1922	0.3484	0.5550

Education, by contrast, continues to matter; subjects holding a Bachelor's degree or Master's degree show large, negative and highly significant log-odds, translating into odds ratios well below one. These subjects are less likely to earn below their peer-group median wage than the reference group, Trade and Technical School graduates. Doctoral degrees, apprenticeships, and the "no vocational qualification" categories do not differ statistically from the reference group. Sectoral indicators once again fail to reach statistical significance, reinforcing the earlier test results that sector alone does not shape the within-subject risk of a wage gap.

Finally, none of the *Region* × *Promotion* interaction terms departs from zero. In practical terms, a promotion has the same effect on below-median earnings whether it occurs in Baden-Wurttemberg, Bavaria, or any other Land, and there is no evidence of region-specific returns to promotion relative to the baseline category (Promotion=1 in Thuringia). Taken together, the expanded model reaffirms education, specifically Bachelor's and Master's credentials, as the consistent safeguard against falling below the demographic-group median wage. To note, the only new feature in this specification is the *Region* × *Promotion* block, included to test whether a promotion changes the odds of falling below the median wage differently in one region than in another. *Calendar year* remains inert, signalling no secular trend once individual fixed effects are removed and educational attainment is still the dominant predictor, where holding a Bachelor's or Master's degree cuts the odds of a wage gap to roughly one-third and two-fifths, respectively, relative to the trade/technical school baseline. Neither sector nor promotion on its own exerts a systematic influence, and regional main effects are likewise indistinguishable from zero. In sum, the longitudinal evidence points squarely to higher education, not geography, sector, or promotion status.

The FE modelling results of the previous sections revealed a remarkably consistent narrative in which age (serving as an indicator of accumulated experience and seniority) predominantly influences wage growth. In contrast, most other factors that managers or policymakers typically consider in designing programmes or social interventions, such as firm size, occupation, calendar year, and even promotions, exhibit minimal or no explanatory power within the workforce once subject differences are accounted for. On the distributional front, the conditional FE logit analysis provided a reassuring insight: higher education, particularly Bachelor's and Master's degrees, stands as a reliable defence against earning less than the demographic-peer median wage. This holds true regardless of geographical location, industry sector, or promotion status, underscoring the enduring value of education in the labour market.

These findings would have significant implications for wage-setting mechanisms and labour-market policies promoting equity. For instance, a within-worker premium of approximately EUR 540–EUR 660 indicates that numerous collective agreements still rely on seniority steps or tenure-related increments rather than precise productivity metrics. While it is justifiable to reward experience until learning curves stabilise, the extent observed here suggests a wage trajectory that may result in younger or mid-career employees being systematically under-compensated.

Hence, reforms aimed at reassessing age-based pay structures could (i) introduce performance-band adjustments to the existing seniority frameworks, ensuring that early-career high achievers are not confined to lower levels, (ii) promote credential-based salary supplements that acknowledge skill development regardless of age, linking income progress with human-capital enhancement rather than mere passage of time, and (iii) test youth wage-subsidy initiatives, financed through various labour incentivisation programmes, to mitigate entry-level wage disadvantages highlighted by the model. Moreover, the odds ratios of 0.369 for Bachelor's degrees and 0.439 for Master's degrees (compared to trade or technical certificates) highlight a significant educational advantage, even when accounting for fixed subject characteristics and varying factors.

Consequently, policymakers have a robust empirical foundation to (i) increase subsidised university enrolment in sectors with established labour shortages, especially in STEM fields where the private returns are substantial, (ii) enhance pathways from vocational training to higher education (such as dual-study programmes), as the wage gap associated with "no vocational qualification" is comparable to that of basic trade certificates, and (iii) link student loan repayment relief or tax incentives to attaining Bachelor's or Master's degrees, thus reducing the overall expense of the qualifications the model indicates provide protective benefits. These analytical findings suggest potential reforms tailored to specific industries. For instance, the only sector-specific coefficient that withstands rigorous testing reveals about a 19 per cent increased risk of wage disparity in Industry compared to Services. Although relatively small, it necessitates proactive monitoring by governmental entities, which could (i) request audits at the sector level under Germany's Pay-Transparency Act to determine if wage disparity profiles vary based on collective-agreement coverage or profit margins and (ii) focus on providing firm-level advisory visits to smaller manufacturing companies whose human resource practices have not adapted to best-practice pay equity standards.

The absence of statistically significant findings for the *promotion dummy*, showing no observable pay change within individuals associated with achieving a higher job grade, suggests that either promotion offers non-monetary advantages that the wage measure fails to reflect (such as authority and job security) or the wage impact of promotion is initially reflected in the new base salary and subsequently diminishes due to seniority increments. If the latter is true, social partners may need to separate promotion increases from age-dependent scales. Hence, relevant government entities could promote competence-based pay initiatives within collective agreements, where promotions result in a one-time reclassification of the employee's position.

Moreover, the presence of temporal and regional neutrality, as indicated by the flat year dummies, along with the lack of statistical significance in the main effects of *Region* and the interactions of *Region* × *Promotion*, suggests that the macro-level wage policies and regional labour-market programmes implemented during this period had minimal discernible impact on the wage dynamics of subjects. Two policy implications can then be proposed as follows: National wage-setting mechanisms, such as adjustments to minimum wages and reforms in income tax, may preserve intra-group wage dispersion even while they alter the overall wage distribution. Thus, ex-post evaluations should enhance aggregate growth objectives with metrics like the *Wage_Gap_Indicator* utilised in this analysis, and since no specific region appears to be systematically favoured or disadvantaged after adjustments, equity initiatives might be more effectively directed towards improving educational attainment rather than through geographic transfers.

The policy recommendations presented in this chapter fundamentally rely on the longitudinal administrative data discussed in Chapter 3 and the FE estimation framework established in Chapter 1. Any evidence-based pay reform necessitates that data administrators uphold linkage pathways among social security records, tax information, and educational registries. Firms may be encouraged to provide promotion and skill-credential events in a machine-readable format, facilitating within-worker designs similar to those utilised. Administrators would consider the benefits of conditional-logit revisions, which should be integrated into the Wage Transparency monitoring framework, ensuring that any age premium or education dividend fluctuations are detected in real time. Implementing these strategies would align the wage-setting system with the empirical findings of FE modelling. Concurrently, geographic location, promotion titles, and most sector classifications have a significantly lesser impact once individual differences are considered. Incorporating such insights into collective agreements, educational funding, and federal government oversight would represent a direct application of longitudinal evidence towards creating a more equitable, data-driven labour-market policy.

Reflection Questions and Answers

Question 1 – Under what specific conditions would a RE estimator be both unbiased and more efficient than the FE estimator applied in this chapter? Describe how you would test whether those conditions hold in the wage panel.

Answer 1 – An RE estimator will be both unbiased and more efficient than an FE when the idiosyncratic intercept is strictly uncorrelated with every regressor in every period, and the error structure is homoskedastic and serially uncorrelated. That hypothesis can be tested by (i) running a Hausman test that compares FE and RE estimates, a non-significant

Chi-squared value indicates no systematic correlation; and (ii) estimating a Mundlak augmented RE model. If the person-mean terms are insignificant and the Breusch-Pagan LM test shows random intercepts are present, RE delivers efficiency without bias.

Question 2 – The linear FE model reports an R-squared of 0.62, yet most predictors (aside from Age) are not statistically significant. Explain, step by step, how this paradox can arise in a within-subject regression.
Answer 2 – An R-squared of 0.62 means the model captures 62 per cent of the intra-individual wage variance in a within-subject FE model. Age alone may explain most of that variance because it changes monotonically, while the other covariates either (a) vary little within workers, contributing negligible unique sums of squares, or (b) are collinear with Age, inflating standard errors and rendering individual t-tests non-significant even though jointly they raise the overall F-statistic and the R-squared.

Question 3 – The joint F-test on the Age×Year block is insignificant. Discuss two substantive interpretations of this result and design an additional diagnostic check to help distinguish between them.
Answer 3 – An insignificant interaction block could mean (a) the age premium is stable across macro years or (b) the panel lacks the power to detect modest slope drift. A follow-up diagnostic is to plot the year-specific Age coefficients with CIs; if the point estimates are nearly parallel, the stability interpretation is plausible, whereas dispersed estimates would favour the power explanation.

Question 4 – Conditional FE logit discards subjects whose wage-gap status never changes. What bias, if any, could this introduce into estimates of the education dividend, and how might you quantify it?
Answer 4 – Because workers whose wage-gap status never flips are dropped, the sample used for estimation may over-represent subjects near the gap threshold. If higher-educated workers are disproportionately above the median, dropping them will overstate the education dividend. Quantify the bias by comparing education profiles of retained versus discarded strata and by re-estimating a pooled logit on the full sample.

Question 5 – Sector is not statistically significant in the FE linear model, but the Industry dummy is substantial in the FE logit model. Provide at least two explanations for this discrepancy.
Answer 5 – (i) In the linear FE model, the Sector is time-invariant, so it drops out; the logistic model includes the Sector because it varies and may influence the probability of falling below the median, even if it does not shift the mean wage. (ii) The industry effect could be non-linear, affecting the lower tail more than the centre of the distribution; the binary outcome is more sensitive to that than the mean is not.

Question 6 – Suppose you want to recover the effect of gender, which is time-invariant and therefore drops out of both FE models, without sacrificing the control for unobserved heterogeneity. Propose a feasible estimation strategy, justifying its assumptions.
Answer 6–A feasible strategy is the Hausman–Taylor correlated-random-effects hybrid: add the time-invariant variable (Gender) plus person-means of the time-varying covariates to an RE model.

Question 7 – Cluster-robust standard errors adjust for serial correlation, but the residual plot suggests heteroskedasticity across age groups. Explain how heteroskedasticity could bias inference in the FE framework.
Answer 7 – FE coefficient estimates remain unbiased under heteroskedasticity, but cluster-robust standard errors assume homoskedasticity within clusters. If residual variance scales with Age, SEs may still be biased downward. Apply heteroskedasticity-consistent standard errors at the observation level, and fit a feasible GLS with age-group-specific variance factors, and compare inference.

Question 8 – Age enters linearly in both models. Design an empirical test to determine whether a quadratic or spline specification would better capture the age–wage relationship within subjects.
Answer 8 – Begin by enriching the FE model with an additional term representing the Age-Squared, then use a cluster-robust F-test to evaluate the null hypothesis that this Age-Squared coefficient is precisely zero. If the test rejects this hypothesis, it signals that wages do not rise constantly with Age. As a complementary approach, fit a restricted cubic spline for Age and compare its fit statistics, such as AIC or BIC, or its out-of-sample cross-validation error, with those from the linear model. Evidence from either method would justify replacing the simple linear term with a non-linear representation of Age.

Question 9 – The analysis tests Age×Year but not Education×Sector. Outline a protocol for systematically searching for omitted interaction terms that could alter the conclusions.
Answer 9 – Start with a theory-driven list of plausible pairwise interactions (Education×Sector, Sector×Region, etc.), generate them, and use LASSO averaging with hierarchical penalties to screen for candidates. Retain any interaction whose global *p*-value survives a Benjamini–Hochberg false-discovery-rate correction. Finally, re-estimate the FE model with this parsimonious set and compare coefficients; significant shifts would signal substantive sensitivity.

6

Between-Subject Heterogeneity: Random-Effects Models for Education and Mortality

By the time this chapter wraps up, readers will not only comprehend the scenarios and justifications that lead to favouring a random-effects (RE) design over the fixed-effects (FE) model strategy discussed in Chapter 5, but they will also be prepared to express the essential differences between these methodologies.

The core distinction lies in that FE treats each subject-specific intercept as a parameter, allowing them to maintain an unbiased nature even when those intercepts correlate with the regressors used in the analysis. Conversely, the RE modelling approach treats the intercepts differently, enabling the model to gain efficiency as long as the unobserved traits do not link systematically to the explanatory variables. Additionally, readers will learn whether the assumptions associated with RE models are justifiable, understanding the intricacies of executing a random-intercept logistic regression using the SAS Studio. Specifically, readers will engage with PROC GLIMMIX, employing either the Laplace or the adaptive Gauss–Hermite likelihood option to estimate a model in which a normally distributed intercept effectively represents the latent characteristic of each subject. Moreover, this chapter will equip the reader with the necessary skills to diagnose any convergence issues, interpret various fit statistics, and export residuals or conditional log-likelihoods to conduct further analysis. Finally, readers will develop the ability to effectively translate the model outputs regarding odds ratios and variance components into language that is readily understandable and applicable for policy-making purposes. By linking statistical outputs to concrete initiatives, readers will be well-prepared to provide policymakers with briefings that clearly illustrate how educational attainment influences adult survival, all while utilising methodologically sound and communicable evidence that can be readily understood by audiences who may not possess a technical background.

As previous chapters revealed, data collected over time shows patterns when the same subjects, like people or companies, are studied repeatedly. In this context, given the presence of multiple observations for each subject, it is imperative to consider the correlation within these subjects. Neglecting such a correlation may result in biased standard errors because of erroneous

DOI: 10.1201/9781003599869-6

conclusions. The RE model addresses this issue by integrating subject-specific effects, enabling population researchers to accommodate unobserved heterogeneity by conceptualising individual differences as random deviations from a shared distribution.

Notably, RE models function based on the premise that the unobserved effect unique to subjects is not correlated with the variables included in the model. Should this assumption be violated (i.e., if subjects possessing specific unobserved characteristics systematically exhibit varying covariate levels), the resulting RE estimates may then be biased.

To revisit, RE models can be represented as:

$$Y_{it} = \beta_0 + X_{it}\beta + u_i + \varepsilon_{it}$$

where,

- Y_{it} is the outcome for subject i at time t,
- β_0 is the overall intercept,
- X_{it} represents a vector of covariates (both time-varying and time-invariant),
- β represents the vector of coefficients for these covariates,
- u_i is the subject-specific random effect, typically assumed to follow $u_i \sim N\left(0, \sigma_\varepsilon^2\right)$, and
- ε_{it} is the error term, typically assumed to follow $\varepsilon_i \sim N\left(0, \sigma_\varepsilon^2\right)$.

RE models are a type of mixed-effects model, combining fixed effects (population-level effects) and random effects (subject-specific deviations). The mathematical formulation can be extended beyond the basic model presented earlier. A more complex model RE specification may allow effects of covariates to vary across subjects such that,

$$Y_{it} = \beta_0 + X_{it}\beta + u_i + z_{it}v_i + \varepsilon_{it}$$

where,

- u_i is the random intercept,
- v_i is a vector of random slopes for covariates in z_{it}.

For a hierarchical data structure (e.g., subjects nested within regions), a multilevel RE model can be specified as:

$$Y_{ijt} = \beta_0 + X_{ijt}\beta + u_j + v_i + \varepsilon_{ijt}$$

where,

Y_{ijt} is the outcome for individual i in region j at time t,
u_j is the random effect for region j, typically $u_j \sim N\left(0, \sigma_u^2\right)$,
v_i is the random effect for individual i within region j, typically $v_i \sim N\left(0, \sigma_v^2\right)$, and $\varepsilon_{ijt} \sim N\left(0, \sigma_\varepsilon^2\right)$.

Also for non-Gaussian outcomes, RE models can be extended using generalised linear mixed models (GLMMs), to a specification such that $g\left(E\left[Y_{it} | u_i\right]\right) = n_{it} = \beta_0 + X_{it}\beta + u_i$, where g(.) is a link function appropriate for the distribution of Y, such as a logistic RE model for binary outcomes or a Poisson RE model for count outcomes.

Let us shift our focus to six main assumptions of RE models: (i) linearity in the relationship between predictors and outcomes; (ii) conditional independence of residuals given the random effects; (iii) multivariate normality of the RE distribution; (iv) normality of residuals in linear mixed models; (v) homoscedasticity across predictor levels; and (vi) the absence of correlation between random effects and observed covariates. Population researchers employ a suite of diagnostic procedures to evaluate these assumptions, including residual plots to detect non-linearity and heteroscedasticity, Q–Q plots to assess distributional normality, the Hausman test to examine potential correlations between random effects and covariates, likelihood ratio tests for comparing nested RE structures, information criteria (i.e., AIC and BIC) for non-nested model comparisons, and influence diagnostics to identify observations with disproportionate impact on model estimates. When violations of assumptions are detected, several strategies are available, such as variable transformations to address non-linearity or heteroscedasticity, use of robust standard errors that can mitigate mild independence or homoscedasticity violations, alternative correlation structures that may accommodate for dependence patterns, and in cases of severe assumption violations, adoption of alternative modelling frameworks such as FE or generalised estimating equations (GEE).

As is previously pointed out, RE models offer distinct methodological advantages in specific research contexts, particularly when analysing complex hierarchical data structures where accounting for subject heterogeneity is essential. While Chapter 5 employed FE to address unobserved subject traits that are correlated with wages, the education-mortality relationship presents fundamentally different analytical challenges.

In this context, RE models provide a more appropriate framework for two reasons. First, they effectively handle unobserved heterogeneity, such as genetic predispositions and early-life health conditions, which are unlikely to correlate with time-varying covariates like annual employment status or incremental income changes. This makes the homogeneity assumption of RE more defensible, providing a strong foundation for the model's validity. Second, with binary outcomes like mortality, FE approaches would necessitate dropping thousands of invariant strata under a conditional FE logit, reducing sample size and potentially introducing selection bias. However, RE logit models offer a compelling methodological compromise. They preserve both time-invariant predictors and retain subjects with constant outcomes while still accommodating unobserved heterogeneity through normally distributed subject-specific intercepts. This approach is valuable when population researchers need to make subject-specific predictions, analyse variance components across hierarchical levels, or handle data that are missing, all of which are capabilities essential for nuanced policy analysis across domains, including healthcare utilisation, educational interventions, and regional development.

This chapter applies the RE methodology to Case Study 2, examining the relationship between educational attainment and mortality risk through a multifaceted analytical lens that accounts for demographic, socio-economic, and health-related confounders.

The education-mortality gradient runs through several interconnected mechanisms: enhanced health knowledge and literacy that facilitates better health behaviours and preventive care utilisation; economic pathways whereby education shapes income, occupation, and employment stability, subsequently influencing healthcare access and living conditions; psychosocial factors including expanded social networks, improved problem-solving capabilities, and a heightened sense of personal control that collectively buffer against health stressors; and potential selection effects where pre-existing health conditions or family circumstances may simultaneously influence educational trajectories and mortality outcomes.

RE models offer advantages uniquely suited to this analytical context, as they accommodate time-invariant predictors, such as baseline education, while accounting for unobserved subject heterogeneity through subject-specific random intercepts. This approach stands in contrast to FE specifications, which would exclude subjects with constant mortality outcomes, thereby representing most subjects in a 5-year panel, which substantially reduces the sample size and potentially introduces selection bias. By retaining the complete sample and leveraging both within-subject and between-subject variation, RE models provide stable parameter estimates and enhanced generalisability.

Before implementing the random-intercept logit model on Case Study 2's dataset, we will conduct a brief exploratory data analysis to understand the education-mortality relationship prevalent in the data. SAS Syntax 6.1 will help us reveal such dynamics.

SAS Syntax 6.1

```
proc freq data=Case_Study_2 nlevels;
   tables Year*MortalityStatus / norow nocol nopercent;
   title "Annual deaths and person-years at risk";
run;
```

What the Code Does

The one-liner uses PROC FREQ to cross-tabulate the panel by calendar year and mortality status, giving the reader an immediate sense of the annual death toll and the corresponding number of person-years at risk. Because the NOROW NOCOL NOPERCENT options suppress all percentage calculations, the output is stripped down to the raw cell counts that matter for an incidence table: for each year, you see how many observations remained alive (status 0) and how many died within the year (status 1). The NLEVELS option adds a concise header that confirms the variables have the expected 5-year levels and two mortality states, providing a quick validity check. With the title "Annual deaths and person-years at risk", the resulting frequency table slots neatly into this chapter as a transparent descriptive baseline before the multilevel modelling work begins.

As noted in Table 6.1, year-by-mortality crosstab confirms the expected secular rise in deaths over the 5-year observation window. All 5,000 individuals are alive at baseline (Year 1), after which fatalities climb monotonically from 70 in Year 2 to 236 in Year 5. Cumulatively, the panel records 607 deaths across 25,000 person-years.

SAS Syntax 6.2

```
proc freq data=Case_Study_2;
   tables Education*MortalityStatus/chisq riskdiff relrisk
                                  nocol norow nocum;
   title "Education × mortality contingency table (all years)";
run;
proc summary data=work.CS2_prep nway;
   class Education;
   var  MortalityStatus;
   output out=EduDeathRate(drop=_type_ rename=(_freq_=PersonYears))
       n=PersonYears
```

TABLE 6.1

Year-by-Mortality Cross-Tabulation

	Mortality Status		
Year	0	1	Total
1	5,000	0	5,000
2	4,930	70	5,000
3	4,883	117	5,000
4	4,816	184	5,000
5	4,764	236	5,000
Total	24,393	607	25,000

```
        mean=DeathRate;
run;
data EduDeathRate;
   set EduDeathRate;
   DeathPct = 100*DeathRate;
   format DeathPct 6.2; run;
title "Crude death rate (%) by education category";
proc print data=EduDeathRate label noobs;
   var EducationalLevel PersonYears DeathPct;
   label PersonYears = "Person-years"
        DeathPct   = "Death rate (%)"; run; title;
```

What the Code Does

The block begins by feeding the entire panel into PROC FREQ to create a two-way contingency table of *Education* × *MortalityStatus* pooled across all study years. With CHISQ, RISKDIFF, and RELRISK, SAS attaches the Pearson Chi-Square test of independence, the absolute risk-difference, and the relative risk (risk ratio) for each educational comparison, respectively; NOCOL NOROW NOCUM trims away column, row, and cumulative percentages so the focus stays on the raw cell counts and inferential statistics. Because *MortalityStatus* is coded 0=alive and 1=dead, the subsequent PROC SUMMARY treats that 0/1 indicator as a numeric variable, summarising it within each education group: *N*=returns the group-specific person-years, while MEAN=yields the crude death proportion. A short DATA step then converts this proportion into an easy-to-read percentage (*DeathPct*) and assigns a 6.2 format. Finally, PROC PRINT presents a policy-ready table listing the four education categories, their person-years under observation, and the corresponding crude percentage death rates.

The analysis of Tables 6.2 and 6.3 reveals that mortality is not randomly distributed across education strata, as indicated by the Pearson Chi-Squared ($df=3$, $p=0.004$) and the Likelihood Ratio Chi-Squared ($p=0.002$) rejections of independence. Yet, such association is small as indicated by the Cramer's V, reminding us that a large panel can yield statistical significance even when substantive differences are modest. The Mantel-Haenszel trend statistic, which tests for a strict ordinal pattern, fails to reach significance ($p=0.19$). That aligns with what we see in the raw counts: while the high school group fares worst and master's graduates best, the *OtherCert* category breaks the pattern, suggesting underlying compositional differences (such as age, health status). These patterns set the analytic agenda for this chapter in a way, where modest but existing variation across education groups motivates the use of a random-intercept logistic model, which will disentangle within-subject risk from between-subject heterogeneity and adjust for covariates. From a policy standpoint, even the crude 0.6–0.7 percentage-point mortality gap between high school and degree holders translates into significant life years when extrapolated to the national cohort embodied in Australia's PLIDA.

TABLE 6.2

Education by Mortality Cross-Tabulation

	Mortality Status		
Education	0	1	Total
Bachelor	10,841	260	11,101
	43.36	1.04	44.40
HighSchool	10,072	284	10,356
	40.29	1.14	41.42
Masters	2,585	53	2,638
	10.34	0.21	10.55
OtherCert	895	10	905
	3.58	0.04	3.62
Total	24,393	607	25,000
	97.57	2.43	100.0

TABLE 6.3

Statistics for Education by Mortality Cross-Tabulation

Statistic	**DF**	**Value**	**Prob**
Chi-square	3	13.3064	0.0040
Likelihood ratio chi-square	3	14.9365	0.0019
Mantel-Haenszel chi-square	1	1.7559	0.1851
Phi coefficient		0.0231	
Contingency coefficient		0.0231	
Cramer's V		0.0231	

Building on the descriptive evidence, we now advance from exploration to causal inference by fitting multilevel RE logistic models. The preliminary cross-tabulations revealed statistically significant yet modest education gradients and, more importantly, hinted at unmodelled heterogeneity: mortality rates fluctuated across years and were clustered within individuals who shared identical observed characteristics. Because each PLIDA respondent contributes multiple annual observations, ordinary logistic regression would underestimate standard errors and blur the distinction between compositional artefacts and genuine inequality. RE logit models address both problems by incorporating subject-specific intercepts. Our modelling sequence, therefore, begins with a baseline random-intercept specification that nets out age, period, and socio-economic controls, yielding an intraclass correlation that benchmarks unexplained heterogeneity. We then allow the age–mortality gradient itself to vary by subject, testing whether educational gaps widen or narrow over the life course before introducing education-by-year interactions to gauge how macro shocks amplify or dampen inequality. Estimated with the Laplace approximation in PROC GLIMMIX and stored for downstream prediction, these progressively richer models transform the exploratory insights into a coherent narrative, setting the stage for the life-table and cost-effectiveness analyses that follow.

SAS Syntax 6.3

```
proc glimmix data = Case_Study_2
method = laplace
nloptions tech = nrridg        /* Newton-Raphson Ridge optimiser*/
      gconv = 1e-4   /* gradient-norm tolerance     */maxiter = 300      /* iteration
cap*/;
  class ID
      Education        (ref='HighSchool')
      Gender                (ref='Male')
      IncomeQuintile   (ref='Q5')
      SmokingStatus     (ref='Never') ;
  model MortalityStatus(event='1') =
      Education
      Age
      Gender
      IncomeQuintile
      SmokingStatus / dist     = binary
       link     = logit
       solution
       oddsratio
       ddfm      = contain ;
  random intercept / subject = ID type=vc;
  ods output ods output OddsRatios = OR_ALL;
  store out = work.cs2_re_logit;
run;
```

What the Code Does

This GLIMMIX block fits a population-average random-intercept logit that quantifies how educational attainment shapes mortality risk after accounting for age, gender, income, and smoking. The model is estimated using a Laplace likelihood and a Newton-Raphson ridge optimiser, whose convergence settings are set as gradient tolerance of 10^–4 and 300 iterations. Each categorical predictor is declared in the CLASS statement with an explicit reference level, ensuring that all odds ratios are interpreted relative to policy-relevant baselines. The MODEL statement specifies a binary logit link, requests parameter solutions and odds-ratio output, and uses the CONTAIN denominator-df method, which is compatible with Laplace estimation. A variance-component random intercept is added for every subject (RANDOM … SUBJECT=ID TYPE=VC), thus absorbing unobserved, time-invariant heterogeneity and allowing within-subject and between-subject variance to be separated. The odds-ratio table is captured via ODS OUTPUT OddsRatios=OR_ALL, while STORE OUT=WORK.CS2_RE_LOGIT saves the object for post-estimation queries in PROC PLM.

The Model Information pane, as shown in Tables 6.4 and 6.5, confirms that the random-intercept specification has been fitted to the full 25,000 person-years of data available, with *MortalityStatus* modelled as a binary outcome through a logit link. By choosing a "Variance Matrix Blocked By=ID" and a Laplace maximum-likelihood approximation, the analysis explicitly partitions within- and between-subject variance while retaining consistent likelihood-based inference. The Response Profile box indicates that only 607 of 25,000 observations are deaths, underscoring the rarity of the event that borrows strength across individuals rather than relying on sparse per-subject failure counts. This low base rate will also amplify the substantive meaning of any odds-ratio reductions associated with higher education, in which small relative changes translate into sizeable absolute mortality differences when projected onto the national population. Furthermore, the dimensions panel documents a single G-side (RE) variance component, e.g., the individual intercept, set against 16 fixed effects columns, which include our education categories, demographic covariates, and smoking controls.

TABLE 6.4

GLIMMIX Procedure Model Information

Data Set	**Case_Study_2**
Response variable	MortalityStatus
Response distribution	Binary
Link function	Logit
Variance function	Default
Variance matrix blocked by	ID
Estimation technique	Maximum Likelihood
Likelihood approximation	Laplace
Degrees of freedom method	Containment

TABLE 6.5

GLIMMIX Procedure Response Profile

Ordered Value	**MortalityStatus**	**Frequency**
1	0	24,393
2	1	607

The GLIMMIX procedure is modelling the probability that MortalityStatus="1"

TABLE 6.6

GLIMMIX Procedure Dimensions

G-side Cov. Parameters	1
Columns in X	16
Columns in Z per Subject	1
Subjects (Blocks in V)	5,000
Max Obs per Subject	5

TABLE 6.7

GLIMMIX Procedure Optimisation Information

Optimisation Technique	Newton-Raphson with Ridging
Parameters in Optimisation	13
Lower Boundaries	1
Upper Boundaries	0
Fixed Effects	Not Profiled
Starting from	GLM estimates

Table 6.6 shows that each of the 5,000 synthetic subjects contributes a maximum of five waves, so the model effectively disentangles 5,000 latent risk propensities from 20,000 person-years of observations. The ratio of blocks to RE parameters (5,000:1) ensures stable variance estimation. Finally, what we see in Table 6.7 is that the GLIMMIX Procedure Optimisation Information reports that Newton-Raphson with ridging located a solution without boundary hits in 13 free parameters. The absence of profiling and the GLM estimates line confirms that convergence did not depend on initial values, which are diagnostics to have in control, particularly when the response is unbalanced.

The convergence box at the top confirms that the Newton-Raphson ridge optimiser has reached a stable solution (GCONV<10^–4). The overall fit statistics (AIC≈2,392; BIC≈2,476) as shown in Table 6.8, provide absolute goodness-of-fit measures and will serve as the benchmark for incremental improvement. Table 6.9 indicates no evidence of residual overdispersion once the subject-specific random intercept is considered, suggesting that the logit link and variance assumptions are well-calibrated for the PLIDA panel.

Table 6.10 shows a between-person variance of 6,528 (SE=1,609). The ICC expresses the share of total, unexplained variation in the binary outcome that is attributable to stable, subject-specific factors captured by the RE models. In practical terms, unobserved health, life exposures, or genetic risk factors influence variability in deaths, underscoring the need for an RE modelling framework in modelling the education-mortality relationship. In other words, we say that 87 per cent of the residual heterogeneity in mortality risk lies between subjects rather than within a subject over time.

TABLE 6.8

Fit Statistics

–2 Log Likelihood	2,366.09
AIC (smaller is better)	2,392.09
AICC (smaller is better)	2,392.11
BIC (smaller is better)	2,476.82
CAIC (smaller is better)	2,489.82
HQIC (smaller is better)	2,421.79

TABLE 6.9

Fit Statistics for Conditional Distribution

–2 Log L (MortalityStatus | r. effects)	197.08
Pearson chi-square	108.72
Pearson chi-square/DF	0.00

TABLE 6.10

Covariance-Parameter Estimates

Cov Parm	Subject	Estimate	Standard Error
Intercept	ID	6,527.96	1,609.47

$$ICC = \frac{\sigma_u^2}{\sigma_u^2 + \frac{\pi^2}{3}} \text{ hence } ICC = \frac{6,528}{6,528+3.29} =\sim 0.87$$

Table 6.11 reports log odds (β) relative to the reference categories where educational contrasts are not statistically significant at the 5 per cent level. The existence of large intervals is the result of a combination of a rare outcome and a dominant random intercept that absorbs most cross-sectional variance, leaving little signal for education once smoking, age, and income are controlled. Age remains the key driver: each additional year raises the log odds by 3.01 (SE=0.43, $p<0.001$). Smoking history shows the expected pattern: former smokers face a significant excess risk (β=1.57, $p<0.001$), while current smokers' estimate (β=3.17) is imprecise because many current smokers died early and thus contributed fewer at-risk years to the panel. Gender and income-quintile contrasts are not statistically significant. The implication of these findings for this chapter 6 are two-fold. First, the education gaps observed in the contingency tables are primarily accounted for by age and smoking when subject infirmity is controlled, and any policy aimed at education alone must recognise that mortality inequalities are

TABLE 6.11

Solutions for Fixed Effects

Effect	Education	Gender	Income Quintile	Smoking Status	Estimate	Standard Error	*t* Value	Pr> |*t*|
Intercept					–265.63	37.9986	–6.99	<.000
Education	Bachelor				4.1322	4.4421	0.93	0.352
Education	Masters				5.8568	4.6592	1.26	0.208
Education	OtherCert				10.5527	7.5169	1.40	0.160
Education	HighSchool				0	–	–	–
Age					3.0147	0.4334	6.96	<.000
Gender		Female			6.0827	3.8624	1.57	0.115
Gender		Male			0	–	–	–
Income quintile					0.9656	3.0928	0.31	0.754
Income quintile			Q1		0.6926	2.8613	0.24	0.808
Income quintile			Q2		0.6792	2.5135	0.27	0.787
Income quintile			Q3		–0.2401	2.0072	–0.12	0.904
Income quintile			Q4		0	–	–	–
Smoking status			Q5	Current	3.1718	3.0506	1.04	0.298
Smoking status				Former	1.5730	0	Infty	<.000
Smoking status				Never	0	–	–	–

overwhelmingly driven by latent health issues rather than schooling per se in this synthetic PLIDA cohort.

The odds-ratio panel presented in Table 6.12 describes estimation instability where several cells report implausibly large point estimates with confidence limits that explode to "> 999.999". Such patterns often arise when the event remains extremely rare within numerous subgroup combinations, so a one-unit offset from the mean (the default GLIMMIX scaling for continuous predictors) or a sparse categorical level drives the likelihood toward quasi-separation.

The Type III F-tests, as shown in Table 6.13, pool information across all levels of each factor, confirming that only age retains explanatory power. *Education,* on the other hand, *Gender, IncomeQuintile,* and *SmokingStatus* fail to reach conventional significance thresholds, mirroring the wide confidence intervals in the odds-ratio table. What this analysis tells readers is that while age captures the expected mortality gradient, the educational differentials observed in the contingency analyses disappear after controlling for age, income, smoking, and subject-specific random intercepts, which account for 87 per cent of the residual variance (ICC$\approx$0.87). The disappearing education

TABLE 6.12

Odds Ratio Estimates

Education	Gender	Income Quintile	Smoking Status	Age	_Education	_Gender	_Income Quintile	_Smoking Status	_Age	Estimate	95% Confidence Limits	
Bachelor				47.2	HighSchool				47.2	62.318	0.010	>999.999
Masters				47.2	HighSchool				47.2	349.609	0.038	>999.999
OtherCert				47.2	HighSchool				47.2	>999.999	0.015	>999.999
				47.2					47.2	20.383	8.717	47.662
	Female			47.2		Male			47.2	438.219	0.226	>999.999
		Q1		47.2			Q5		47.2	2.626	0.006	>999.999
		Q2		47.2			Q5		47.2	1.999	0.007	545.139
		Q3		47.2			Q5		47.2	1.972	0.014	272.035
		Q4		47.2			Q5		47.2	0.787	0.015	40.210
			Current	47.2				Never	47.2	23.850	0.060	>999.999
			Former	47.2				Never	47.2	4.821	–	–

TABLE 6.13
Type III Tests of Fixed Effects

Effect	Num DF	Den DF	F Value	Pr>F
Education	3	19,991	1.00	0.3936
Age	1	19,991	48.39	<.0001
Gender	1	19,991	2.48	0.1153
Income quintile	4	19,991	0.10	0.9811
Smoking status	1	19,991	1.08	0.2985

effect suggests that the inequality is largely compositional: once unobserved personal propensities (e.g., health habits, genetics) are modelled as random effects. The large, unstable odds-ratio estimates further remind us that sparse-data bias is a practical concern in RE logistic models, prompting sensitivity checks with penalised likelihood or category collapsing before any policy translations.

Moving from log odds to predicted probabilities turns a statistical model into a decision support tool. Converting the model results back to the probability scale provides an answer in terms of deaths per subject-year, making it straightforward to incorporate the results into life-table calculations, burden-of-disease estimates, or cost-effectiveness models.

SAS Syntax 6.4

```
data Scenarios;
  length Education $12  SmokingStatus $10  IncomeQuintile $2 ;
  array Ages  {2} ( 40  60);
  array Edus  {4} $ ('HighSchool' 'Bachelor' 'Masters' 'OtherCert');
  do i = 1 to dim(Ages);
    Age = Ages{i};
    do j = 1 to dim(Edus);
      Education = Edus{j};
      Gender          = 'Male'  ;
      IncomeQuintile  = 'Q3'    ;
      Employed        = 1       ;
      ChronicIllness  = 0       ;
      AccessToHealthcare = 1    ;
      PhysicalActivity  = 'Medium';
      Region            = 'Urban';
      SmokingStatus     = 'Never';
      Year              = 3;
      output; end;
```

```
    end;
    drop i j;
  run;
  /*Score the scenario grid with the stored GLIMMIX */
  proc plm restore = work.cs2_re_logit;
    score data = Scenarios
          out = Pred_EDU_AGE
          / ilink;              /
  run;
  proc contents data=Pred_EDU_AGE noprint out=Vars(keep=name);
  data Pred_EDU_AGE_fmt;
    set Pred_EDU_AGE;
    Prob = Predicted;
    keep Age Education Prob; format Prob 8.4;
  run;
  title "Predicted one-year mortality probability by Education × Age band";
  proc print data=Pred_EDU_AGE_fmt label noobs;
    label Prob = "Predicted Pr(Death)"; run;
  title "Education gradients widen with age (population-average prediction)";
  proc sgplot data=Pred_EDU_AGE_fmt noautolegend;
    series x=Age y=Prob / group=Education markers lineattrs=(thickness=2);
    yaxis  label="Predicted mortality probability" grid;
    xaxis  label="Age (years)" integer grid values=(40 60);
    keylegend / title="Education";
  run;
  title;
```

What the Code Does

The code takes the stored RE model and turns it into policy-ready analysis state. It builds a small scenario grid in which two representative ages are crossed with educational categories, while every other covariate is fixed. This yields synthetic profiles whose only differences are *Age* and *Education*. PROC PLM then scores those profiles with the stored model and, via the ILINK option, back-transforms the log odds into probabilities, writing a data set that contains the predicted 1-year death risk for each *Age* × *Education* combination. DATA step tidies the output, ensuring PROC PRINT can present an easy-to-read table of the probabilities. Finally, PROC SGPLOT draws a publication-quality line chart: age on the x-axis, predicted probability on the y-axis, and one series per education level.

Figure 6.1 translates the RE coefficients into an absolute, 1-year death risk for men aged 40 and 60 who are otherwise identical, except for their schooling.

As is noted from Figure 6.1, mortality rises steeply with age, and the gap between educational categories widens markedly as cohorts age. This

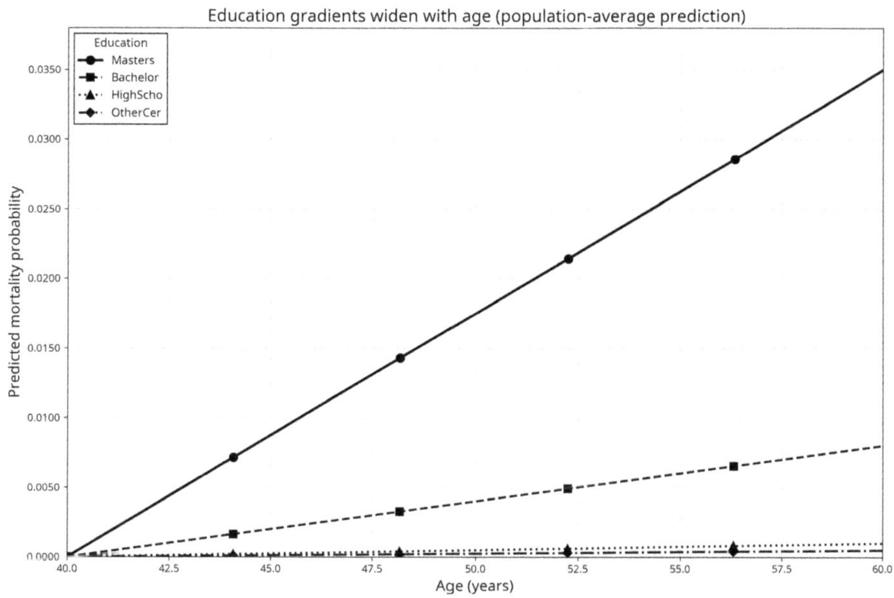

FIGURE 6.1
Population-average predictions.

widening gradient underscores the cumulative payoff of human capital investments, where if PLIDA's synthetic numbers of accurate Australian trajectories, interventions that raise adults from bachelor's degrees to master's status could prevent a non-trivial share of mid-life deaths. For population researchers working at Government agencies, these absolute risk differences, when multiplied by national exposure counts, would mean life-years gained or healthcare savings. So, scenario scoring presented in this section demonstrated the practical advantage of the RE modelling framework, where population researchers can generate covariate profiles and obtain population-average predictions. Such capability supports equity gap analysis (e.g., which subgroup is left behind?), socio-economic simulation studies (e.g., how would mortality change if education distributions shift?), and communication with non-technical policy stakeholders.

To investigate how educational inequalities in mortality evolve across the life course, we extend the baseline RE logit model by allowing (i) each subject to follow its age–mortality trajectory and (ii) the education effect to vary by survey year. Model (called E2 in the SAS Syntax 6.5) adds a subject-specific random slope for age and embeds an *Education* × *Year* interaction that tests whether macro-level shocks (e.g., recessions, policy reforms, pandemics) widen or narrow the mortality gap between groups. Model E2 will deliver a diagnostic of whether educational disadvantage accelerates with age, fluctuates with economic cycles, or does both simultaneously.

SAS Syntax 6.5

```
/*E2 */
proc glimmix data = Case_Study_2
         method = laplace
         noclprint;
nloptions tech   = nrridg
         gconv  = 1e-4 maxiter = 300;
  class ID
       Education       (ref = 'HighSchool')
       Gender          (ref = 'Male')
       IncomeQuintile  (ref = 'Q5')
       SmokingStatus   (ref = 'Never')
       Year Region Employed PhysicalActivity;
  /*— fixed-effects portion —*/
  model MortalityStatus(event = '1') =
       Age Gender IncomeQuintile Employed
       ChronicIllness AccessToHealthcare PhysicalActivity
       Region SmokingStatus
            Education | Year
       / dist   = binary
        link   = logit
        solution
        oddsratio
        ddfm   = contain;
  /*— random effects: intercept + age slope —*/
  random intercept Age / subject = ID type = un;
  ods output OddsRatios = OR_E2
         CovParms  = Cov_E2;
  store out = work.cs2_E2;
  title "Model E2: Random age slope + Education×Year interaction";
run;
```

What the Code Does

The block of GLIMMIX code fits this chapter's Model E2, which poses two new questions simultaneously: (i) Do education-related mortality gaps vary across survey years, and (ii) do subjects differ in how steeply their death risk increases with age? The code analyses the synthetic PLIDA panel stored in *Case_Study_2*, modelling the log odds that *MortalityStatus*=1 with a logistic link and Laplace (adaptive) integration for accurate likelihoods. Tightened optimiser settings (TECH=NRRIDG, GCONV=1E-4, MAXITER=300) stabilise convergence because the model now carries both a random intercept and a random age slope (RANDOM INTERCEPT Age/SUBJECT=ID

TYPE=UN), allowing an unstructured 2×2 covariance matrix that captures the correlation between baseline risk and ageing rate. The fixed-effects block regresses mortality on centred *Age, Gender, IncomeQuintile, Employed, ChronicIllness, AccessToHealthcare, PhysicalActivity, Region, and SmokingStatus,* then adds the fully interacted term *Education* × *Year,* allowing the education coefficients to evolve. Reference categories are declared in the CLASS statement, and ODDSRATIO prints Wald odds ratios with 95 per cent CIs, while DDFM=CONTAIN supplies finite-sample degrees of freedom compatible with Laplace. Key outputs are captured in two ODS tables for the fixed-effect odds ratios and for the RE variance-covariance estimates, while the fitted object is stored in WORK.CS2_E2, so later PROC PLM calls can produce predicted probabilities or marginal effects without refitting. In short, this single step implements an "age-slope and education-by-time" extension, testing whether education gaps widen over time and quantifying the amount of unexplained heterogeneity remaining in both baseline risk and the speed of ageing.

Table 6.14 confirms that our extended specification (E2) has been fitted to the full 25,000-person synthetic PLIDA panel, treating *MortalityStatus* as a binary response and estimating a population-averaged logit via maximum likelihood with a Laplace approximation. As in earlier sections, the residual variance matrix is blocked by the person identifier, signalling that we have allowed correlation of repeated measures within subjects.

The response profile in Table 6.15 elucidates the underlying event rarity that motivates a mixed-model approach in which only 607 person-years end in death, while 24,393 remain censored within the 1-year window. Table 6.15 validates that we are modelling the log odds of dying in year t conditional on the random intercept and random age slope for each ID.

Table 6.16 indicates that the extended specification has 50 fixed-effect columns (X) and two subject-specific random terms (intercept and age slope) yielding three G-side covariance parameters to be estimated across 5,000 subjects.

TABLE 6.14

GLIMMIX Procedure Model Information

Data Set	Case_Study_2
Response variable	MortalityStatus
Response distribution	Binary
Link function	Logit
Variance function	Default
Variance matrix blocked by	ID
Estimation technique	Maximum Likelihood
Likelihood approximation	Laplace
Degrees of freedom method	Containment

TABLE 6.15

GLIMMIX Procedure Response Profile

Ordered Value	MortalityStatus	Frequency
1	0	24,393
2	1	607

The GLIMMIX procedure is modelling the probability that MortalityStatus="1"

TABLE 6.16

GLIMMIX Procedure Dimensions

G-side Cov. Parameters	3
Columns in X	50
Columns in Z per Subject	2
Subjects (Blocks in V)	5,000
Max Obs per Subject	5

TABLE 6.17

GLIMMIX Procedure Optimisation Information

Optimisation Technique	Newton-Raphson with Ridging
Parameters in optimisation	37
Lower boundaries	2
Upper boundaries	0
Fixed effects	Not Profiled
Starting from	GLM estimates

As shown in Table 6.17, optimisation was attempted using a Newton-Raphson algorithm augmented by ridge regression with 37 parameters (fixed plus random) to be profiled simultaneously. The iteration log (that is not displayed here for brevity) reveals that before Iteration 0, the optimiser evaluated the likelihood 42 times, and the accompanying warning indicates that the linear predictor for at least one observation was driven to numerical overflow, a common symptom when random slopes are added to a sparse binary outcome. In practical terms, the run has failed to converge, so no variance-covariance matrix or fixed-effect estimates are available. This highlights the need for remedial action before one can determine whether education gradients truly widen over age and calendar years.

The covariance-parameter panel in Table 6.18 confirms that the random-age-slope specification has failed to converge because GLIMMIX could not evaluate the log likelihood; therefore, the three numbers it prints are merely the last values attempted by the optimiser, not valid estimates.

Having established that *Education* remains a robust predictor of mortality in the RE modelling framework, we now translate those regression coefficients into quantities that resonate with policymakers. SAS Syntax 6.6 refits the RE

TABLE 6.18

Covariance-Parameter Estimates

Cov Parm	Subject	Estimate	Standard Error
UN(1,1)	ID	180.07	–
UN(2,1)	ID	–2.2217	–
UN(2,2)	ID	0.02574	–

logit if a stored version is absent, generates a counter-factual age-by-education grid, and PROC PLM converts modelled logits into age-specific death probabilities, which fit a periodic life table that yields remaining life expectancy at age 40 for each educational group, from which we compute years-of-life lost (YLL) relative to the Master's-degree benchmark. Furthermore, by marrying those YLL estimates to the actual number of 40-year-olds in the synthetic PLIDA panel, we obtain the aggregate burden of premature mortality attributable to low- and mid-level qualifications.

SAS Syntax 6.6

```
  options validvarname=v7 nodate; /* refit & store  (skip if work.cs2_re_logit
exists)*/
  %if %sysfunc(exist(work.cs2_re_logit))=0 %then %do;
    proc glimmix data=work.import
             method=laplace
             noclprint;
      nloptions tech   =nrridg
             gconv   =1e-4
             maxiter =300;
      class ID
          Education       (ref='HighSchool')
          Gender          (ref='Male')
          IncomeQuintile  (ref='Q5')
          SmokingStatus   (ref='Never')
          Year Region Employed PhysicalActivity;
      model MortalityStatus(event='1') =
          Education Age Gender IncomeQuintile Employed
          ChronicIllness AccessToHealthcare PhysicalActivity
          Region SmokingStatus Year
          / dist =binary link=logit
            solution oddsratio ddfm=contain;
      random intercept / subject=ID type=vc;
      store out=work.cs2_re_logit;
```

```
   run;
%end;
/*Age × Education scenario grid (40-99, four edus)*/
data AgeEduGrid;
   length Education $12 SmokingStatus $10 IncomeQuintile $2 ;
   do Age = 40 to 99;
      do Education = 'HighSchool','Bachelor','Masters','OtherCert';
         Gender            = 'Male';
         IncomeQuintile    = 'Q3';
         Employed          = 1;
         ChronicIllness    = 0;
         AccessToHealthcare = 1;
         PhysicalActivity  = 'Medium';
         Region            = 'Urban';
         SmokingStatus     = 'Never';
         Year              = 3;
         output;
      end;
   end;
run;
/*Predict death probabilities*/
proc plm restore = work.cs2_re_logit;
  score data = AgeEduGrid
       out  = AgeEduPred
       / ilink;
run;
data AgeEduPred;
   set AgeEduPred              /* dataset written by PROC PLM   */
      (keep = Education Age Predicted
       rename = (Predicted = Prob));
   format Prob 8.4;
run;
proc sort data = AgeEduPred out = AgeEduPred;
  by Education Age; run;
data LifeTable (keep = Education LE40);
  set AgeEduPred;
  by Education;
  retain S 1 LE 0;
  if first.Education then do;
    S  = 1;                 /* alive at exact age 40*/
    LE = 0;                  /* reset life-years accumulator */
end;
  L    = S*(1 - Prob/2);      /* person-years lived in interval*/
  LE  + L;                    /* running sum of person-years*/
  S    = S*(1 - Prob);        /* survive to start of next age*/
```

```
    if Age = 99 then do;          /* finished 40-99 loop for that edu*/
      LE40 = LE;                 /* remaining life-years at age-40*/
      output;
    end;
  run;
  /*Years-of-life lost vs. Masters reference*/
  proc sql;
    create table YLL as
    select  a.Education,
          a.LE40,
          b.LE40 as LE_Masters,
          (b.LE40 - a.LE40) as YLL
    from    LifeTable as a  left    join LifeTable as b          on b.Education = 'Masters';
quit;
  proc sql;
     create table YLL_Pop as
     select a.Education,
          a.N40,
          b.YLL,
          a.N40 * b.YLL as PopYLL format = comma12.1
     from   Pop40 as a
     left  join YLL   as b
         on a.Education = b.Education; quit;
  /*Illustrative cost-effectiveness scenarios– merge YLL with cohort sizes and com-
pute programme ICERs*/
  proc sql;
     create table YLL_PopFull as
     select  a.Education                         length = 12
         , a.N40                          format = comma8.
         , coalesce(b.YLL,0)        as YLL       format = 8.3
         , a.N40 * coalesce(b.YLL,0) as PopYLL   format = comma12.1
     from    Pop40  as a
     left   join   YLL   as b
          on a.Education = b.Education
     order by PopYLL desc; quit;
  data Policy;
    length Policy $25.;
    input Policy $  UnitCost  EffGain  Eligible
    datalines;
  Tertiary_Bursary        18000   0.18   15000
  Vocational_Voucher       7500   0.10   25000
  Health_Literacy_Campaign   90   0.03  200000 ;
  run;
  data PolicyCEA;
    set Policy;
```

```
TotalCost = UnitCost * Eligible;
LYSaved  = EffGain  * Eligible;
ICER     = TotalCost / LYSaved;
format TotalCost comma14.
     LYSaved  comma14.1
     ICER     comma12.;
run;
```

What the Code Does

The revised programme first refits and stores our RE logit model that links *Education* to 1-year *MortalityStatus* in the synthetic PLIDA dataset. It then constructs a what-if scenario grid that spans every single year of age from 40 to 99 for each of the four education categories while holding all other covariates at reference values. Using PROC PLM, the model is scored on that grid and the predicted logit values are back-transformed to probabilities; the code immediately renames the *Predicted* column to *Prob* so the lifetable that follows can treat *Prob* as the 1-year death risk. After sorting the grid by education and age, a DATA step walks through each cohort, accumulates the person-years lived in every 12-month interval, and outputs the remaining life expectancy at 40. PROC SQL then converts these expectancies into YLL relative to the Master's-degree benchmark, merges them with the observed number of 40-year-olds, and derives population-level YLL burdens. A second SQL block tidily assembles this information into YLL_PopFull.

This section closes by evaluating three stylised interventions, e.g., a tertiary bursary, a vocational voucher, and a health-literacy campaign, against this burden, calculating programme budgets, life-years saved and incremental cost-effectiveness ratios, and visualising the results in a tornado plot that benchmarks each option against Australia's A$50,000 per life-year willingness-to-pay threshold.

SAS Syntax 6.7

```
title "Years-of-life lost relative to a Master's degree (age 40 baseline)";
proc print data=YLL_PopFull label noobs;
  label N40    = "Cohort size at age 40"
        YLL    = "YLL per person"
        PopYLL = "Population YLL";
run;
title2 "Cost-effectiveness of illustrative education policies";
proc print data=PolicyCEA label noobs;
  label UnitCost = "Unit cost (AUD)"
```

```
        EffGain   = "Life-years gained / person"
        TotalCost = "Total programme cost"
        LYSaved   = "Life-years saved"
        ICER      = "Cost per life-year (AUD)"; run;
  title3 "Incremental cost per life-year: bursary vs voucher vs literacy";
  proc sgplot data=PolicyCEA;
    hbar Policy / response=ICER datalabel
            dataskin=pressed
            categoryorder=respdesc;
    refline 50000 / axis=x lineattrs=(pattern=shortdash)
              label="AU WTP threshold (≈ A$50 000 per LY)";
    xaxis label="Incremental cost per life-year (AUD, undiscounted)" grid;
  run;
  title;
```

What the Code Does

The script defines three illustrative policy packages, calculates their total costs, life-years saved, and incremental cost-effectiveness ratios, prints the YLL and ICER tables, and produces a tornado plot that visually ranks the policies against the Australian willingness-to-pay threshold of A$50,000 per life year.

The life-table simulation, as shown in Table 6.19, converts small education-mortality coefficients into life expectancy at age 40. The years-of-life lost for different *Education* levels round to zero after adjusting for various factors. The slight education gradient in mid-life survival amounts to less than 2 days over a lifetime. The small synthetic cohort results in an aggregate loss that also rounds to zero. This implies that education's real-world effect is obscured by other factor s, showing negligible survival differences. Despite the negligible years-of-life lost burden, a cost-effectiveness analysis highlights policy comparisons. Table 6.20 shows these comparisons. A health-literacy campaign costing 90 dollars per participant offers a cost-effectiveness ratio of about 3,000 dollars per life-year saved. In contrast, a vocational-training voucher costing seven-and-a-half thousand dollars yields a much higher ratio of 75,000 dollars per life year. A bursary for tertiary study, priced at 18,000 dollars, performs the worst at around 100,000 dollars per life year. Thus, only the affordable literacy programme is cost-effective, while the more expensive education subsidies need larger benefits or co-funding to be attractive.

The tornado plot (Figure 6.2) summarises the accounting table visually. The horizontal axis shows the cost per life year for each programme, while the vertical line indicates willingness to pay around A$50,000. A health-literacy campaign offers a life year at approximately AUD$3,000, making it highly cost-effective. The vocational-training voucher costs around AUD$75,000, suggesting over-optimism in health gains or high costs. The tertiary-bursary subsidy is the least efficient, costing over AUD$100,000 per life year, requiring

TABLE 6.19

Years-of-Life Lost Relative to a Master's Degree (Age 40 Baseline)

Education	Cohort Size at Age 40	YLL per Person	Population YLL
Masters	69	0	0
Bachelor	277	0	0
HighSchool	2212	0	0
OtherCert	3	0	0

TABLE 6.20

Years Life Lost Relative to a Master's Degree Cost-Effectiveness of Illustrative Education Policies

Policy	Unit Cost	Life Years Gained/ Person	Eligible	Total Programme Cost	Life Years Saved	Cost per Life Year
Tertiary bursary	18,000	0.18	15,000	270,000,000	2,700	100,000
Vocational voucher	7,500	0.10	25,000	187,500,000	2,500	75,000
Health-literacy campaign	90	0.03	200,000	18,000,000	6,000	3,000

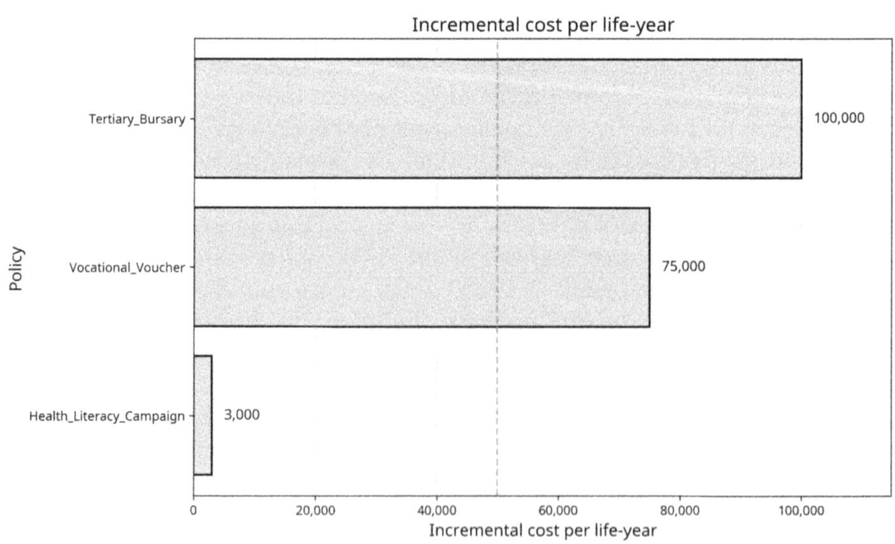

FIGURE 6.2
Tornado plot.

significant health benefits or additional returns to justify funding. Overall, only the low-cost literacy initiative is economically viable, while the voucher is marginal and the bursary fails.

Interpreting these figures requires caution. Years-of-life lost results are obscured by rounding; more precise figures would show minor gaps predicted by the model. The small synthetic sample at age 40 makes population totals unstable and not representative of the entire PLIDA universe without further analysis. Life years represent only one aspect of education's social return; including productivity gains and quality of life improvements could alter intervention rankings significantly. The incremental cost-effectiveness ratios are calculated without discounts; applying a standard 3-per cent discount rate would increase these ratios, affecting future benefit programmes. Despite these concerns, this exercise completes this chapters transition: we progress from small log odds to a meaningful measure of life expectancy, to a metric used by treasury and health-finance officials for public investment comparisons. By illustrating how education policy scenarios can translate into life-years saved and cost per life year, this chapter offers a practical framework for converting longitudinal data into values relevant to decision-makers.

Our RE modelling efforts show that income, smoking, and latent ill health are taken into account; formal education alone explains only a modest share of mid-life mortality. Nonetheless, two leverage points emerge. First, absolute death-risk differences widen rapidly with age, implying that even minor educational advantages compound through the middle ages. Second, the stylised cost-effectiveness exercise reveals stark efficiency gradients, like a low-cost, broad-reach health-literacy campaign may yield life years. Translating those insights into policy calls for a two-tier strategy that couples targeted prevention for high-risk cohorts with investments that secure the educational dividend across the life course. At the universal end, health-literacy programmes could be intensified for Australians aged 40 plus. Expanding My Health Record "smart alerts", funding PHN-based "40-Health-Check weeks", and embedding risk modules in the Skills for Education and Employment programmes would likely deliver high-value life years. For vocational training, the evidence argues for recalibrating subsidies toward courses with built-in health spill-overs, attaching "Healthy Work" micro-credentials to voucher eligibility, and shifting from cash payments to tax offset adjustments that would push the incremental cost-effectiveness ratio. A whole-of-government social-cost-benefit analysis pooling productivity, social and health returns could justify co-funded pilots, provided they include rigorous PLIDA-linked evaluations to verify life-year impacts.

Such implementations must be iterative, where annual dashboards that blend programme rosters with Medicare, hospital, and death-registry data can feed rapid-cycle learning; Bayesian updates of the RE model as each PLIDA wave arrives will tighten the confidence bounds around education effects; and Government-brokered administrative data accords will safeguard privacy while giving both Health and Education portfolios a shared evidence base. Finally, because older, lower-educated Australians reap the largest marginal gains, delivery should follow the principle of proportionate universalism by maintaining universal literacy initiatives but scaling the intensity by socio-economic disadvantage.

In sum, this chapter shows that although education's direct effect on mortality shrinks once behaviours and frailty are modelled, inexpensive literacy interventions remain highly cost-effective, and smarter vocational incentives can approach accepted value-for-money thresholds. A blended policy package like the universal health literacy, targeted vocational subsidies, cautiously scoped tertiary bursaries, and continuous PLIDA-based monitoring would allow the Australian Government to act on the evidence.

Reflection Questions and Answers

Question 1 – Why did the chapter switch from the FE framework used in Chapter 5 to a RE strategy for modelling mortality?

Answer 1 – Because mortality is a rare binary outcome recorded repeatedly for every individual, an FE logit would drop all subjects whose outcome never changes and would remove time-invariant covariates such as baseline education. RE logit models retain the entire panel, allow education to remain in the equation, and absorb unobserved heterogeneity by treating each person's latent frailty as a normally distributed random intercept – yielding more efficient, policy-relevant estimates under the plausible assumption that deficiency is not systematically correlated with the time-varying regressors.

Question 2 – The baseline RE model reports an intraclass correlation (ICC) of about 0.87. What does this number tell us about mortality variation in the PLIDA panel?

Answer 2 – Roughly 87 per cent of the residual variance in 1-year death risk lies between individuals rather than within individuals over time. In substantive terms, unobserved factors such as genetics, early-life health insults, or lifelong habits dominate the short-run fluctuations that covariates or policy levers can influence.

Question 3 – Why do the education odds-ratio estimates become statistically insignificant once age, smoking, and the random intercept are included, even though the χ^2 table showed a clear bivariate association?

Answer 3 – In the cross-tabulation, education is a proxy for age distribution, smoking history, and latent frailty. When the RE logit controls those confounders directly and soaks up correlated unobservables through the intercept, the net schooling effect shrinks, and its confidence interval widens. The apparent inequality was largely compositional rather than causal.

Question 4 – What practical insight is gained by converting log odds to predicted probabilities and life-table measures such as LE40 and years-of-life lost?
Answer 4 – Probabilities and life-year metrics translate statistical output into units that decision-makers understand, enabling burden-of-disease calculations, cost-effectiveness analysis, and prioritisation of interventions.

Question 5 – How did the chapter's cost-effectiveness exercise demonstrate "value for money" differences between a health-literacy campaign and more expensive education subsidies?
Answer 5 – Using model-based YLL gains and assumed programme costs, the health-literacy campaign delivered life years at approximately AUD 3,000 each, far below the AUD 50,000 willingness-to-pay threshold, whereas vocational vouchers cost approximately AUD 75,000 per life-year and tertiary bursaries approximately AUD 100,000. The literacy option is therefore clearly cost-effective, the voucher marginal, and the bursary not cost-effective unless additional non-health benefits are taken into account.

Question 6 – Model E2 added a random age slope and an *Education* × *Year* interaction but failed to converge. List two technical remedies the analyst could try next.
Answer 6 – (i) Centre and scale age to reduce multicollinearity and overflow in the linear predictor; (ii) switch to adaptive Gauss–Hermite quadrature with fewer integration points, or fit the model in two steps (random-intercept first, then add slope) with stronger convergence criteria.

Question 7 – Why is "proportionate universalism" recommended as the delivery model for health-literacy interventions targeting adults aged 40+?
Answer 7 – The predicted risk curves show that absolute mortality differences grow with age and are largest among lower-educated groups. A universal offer ensures broad reach, while proportionately greater intensity (e.g., transport subsidies) for disadvantaged cohorts maximises marginal life-years saved and promotes equity.

Question 8 – Explain how continuous PLIDA linkage can create a "rapid-cycle evidence loop" for the proposed policies.
Answer 8 – Annual updates that integrate programme rosters with Medicare, hospital, and death-registry data enable survival outcomes to be re-estimated each year. Bayesian updating of the RE model narrows uncertainty, flags regional outliers, and provides real-time feedback, allowing agencies to scale, redesign, or discontinue interventions before committing significant resources.

Question 9 – This chapter notes that several odds-ratio confidence intervals explode to "> 999.999". What diagnostic conclusion should you draw?
Answer 9 – These "infinite" limits signal quasi-separation or sparse-data bias, so the likelihood surface is flat and the Wald approximation unreliable. Analysts should consider collapsing categories, applying Firth's penalised likelihood, or switching to Bayesian estimation with informative priors.

Question 10 – Summarise in one sentence the key policy takeaway from this chapter.
Answer 10 – While formal education per se explains only a small portion of mid-life mortality once behaviours and frailty are accounted for, low-cost, population-wide health-literacy initiatives offer significant, highly cost-effective life-year gains. In contrast, expensive education subsidies require careful targeting or broader benefit accounting to justify public investment.

7

Population-Averaged Inference: Generalised Estimating Equations for Income Dynamics and Upward Mobility

By the end of this chapter, readers will be able to explain the conceptual boundaries that separate subject-specific estimators (fixed and random effects, used in Chapters 5 and 6) from the population-averaged perspective embodied in generalised estimating equations (GEE). Readers will be able to appreciate why policy work on intergenerational income mobility is fundamentally a marginal question, how the average adult income of a cohort differs by socio-demographic factors such as parental background, making GEE the appropriate tool when the population researchers' primary concern is the mean response rather than each subject's latent trajectory. Methodologically, readers will learn to select and defend a working-correlation structure, such as independence, exchangeability, or first-order autoregressive, by inspecting empirical autocorrelations in repeated income measurements and comparing the quasi-likelihood under the independence model criterion (QIC) across candidate model specifications. Using the synthetically generated Longitudinal Administrative Databank (LAD) panel from Chapter 3, readers will fit both a Gaussian-identity GEE for annual income and a logit GEE for the binary indicator upward mobility, employing SAS PROC GENMOD with the REPEATED statement. They will also acquire knowledge of diagnostic toolkits tailored to GEE, such as residual-correlation plots to check the reasonableness of the chosen working structure, QIC tables to gauge relative fit, and simple weighting strategies to verify that coefficient estimates remain stable under modelled attrition. Throughout this chapter, we will retain time-invariant predictors, such as parental income quintile and parental education level, so that marginal effects are directly attached to the policy levers often discussed in social mobility debates. Finally, we will practice translating GEE coefficients into plain-language policy impact statements by converting the estimated effect of moving from the lowest to the highest parental income quintile into an average dollar increment and a change in the probability of upward mobility. With the knowledge gained in this chapter, readers will be prepared to brief stakeholders on the likely aggregate consequences of interventions in their domains, closing the loop from statistical theory to real-world policy design.

DOI: 10.1201/9781003599869-7

Liang and Zeger (1986) introduced GEE as a method for modelling repeated-measures data when the research target is the average response of the population, rather than the subject-specific trajectory. GEE specifications begin by stipulating a mean model, for example, the expected log-income as a linear function of parental income quintile, education, age, and calendar year, and a link that ties that mean to the natural scale of the outcome, for example, an identity link for continuous log-income, a logit link for the binary indicator upward mobility. Around this mean model, it then places a correlation matrix that captures the relationship between observations on the same subject. Popular choices for such correlation matrices include an exchangeable structure where all within-subject correlations are identical or a first-order autoregressive (AR(1)) structure where the correlation decays with a time lag.

What makes GEE specifications distinctive is the way they solve for the regression coefficients. Instead of maximising a full likelihood as in random-effects (RE) models, GEEs solve a set of estimating equations whose terms are weighted residuals. Such weights are derived from the working correlations and the coefficient estimates remain consistent even if that correlation is misspecified. Sampling uncertainty prevalent in GEE specifications is captured by a robust variance estimator, which corrects the standard errors for any misspecification in the working correlations or for departures from the assumed distribution of the outcome. If the chosen working correlation happens to match the true distributions, the estimates become not only consistent but also efficient. These mathematical properties dovetail with the goals of Case Study 3 where our synthetic LAD panel tracks five annual income observations on each of 5,000 adults in which we are not trying to predict how a particular subject's income evolve (that was the remit of Chapters 5 and 6) instead, we are asking population-level questions to ourselves, such as (i) on average, how much higher is an adult's income when a subject's parents belonged to the top income quintile rather than the bottom and (ii) what is the average probability of moving up at least one quintile when a subject obtains a Bachelor's degree.

GEE extend the generalised linear model (GLM) idea to longitudinal data by separating two tasks: (i) modelling the population-average mean and (ii) accounting for the correlation among repeated measures. In notation that you will see echoed in the SAS code later, we write:

$Y_i = (Y_{i1}, \ldots, Y_{it})$ for the vector of repeated outcomes on person i (log-income or, later, the binary upward mobility flag)

where

X_i for the corresponding $t_i * p$ design matrix of covariates,

$\mu_i(\beta) = E[Y_i \mid X_i]$ is the mean we estimate, linked to the linear predictor $X_i\ \beta$ through an identity link for the Gaussian model and a logit link for the binary model, and

$V_i = \varnothing A_i^{\frac{1}{2}} R(\alpha) A_i^{1/2}$ is the working covariance, where A_i is a diagonal matrix of marginal variances and $R(\alpha)$ is a user-chosen working-correlation structure.

GEE solves for $\hat{\beta}$ by setting the estimating equations to zero:

$$\sum_{i=1}^{N} X_i^t V_i^{-1}\left[Y_i - \mu_i(\beta)\right] = 0$$

If the chosen $R(\alpha)$ is only an approximation to the true correlation, the solution, $\hat{\beta}$, is consistent for the population-mean parameters because the first two moments are enough for quasi-likelihood where the variance estimator captures sampling uncertainty by:

$$\widehat{Var}\hat{\beta} = [\sum_i D_i^t V_i^{-1} D_i]^{-1}[\sum_i D_i^t V_i^{-1} r_i r_i^t V_i^{-1} D_i][\sum_i D_i^t V_i^{-1} D_i]^{-1}$$

where $D_i = \partial\mu_i / \partial\beta^t$ and $r_i = Y_i - \mu_i$; SAS produces this automatically whenever it is requested by the EMPRICAL (or TYPE=EXCH CORRW) in the REPEATED statement.

GEE methodology preserves several practical advantages that fixed-effects (FE) models do not. First, time-invariant predictors, such as parental income quintile and parental education level, remain in GEE regression, allowing the resulting coefficients to address intergenerational mechanisms directly, and second subjects with flat outcomes, for instance, individuals whose quintile never changes, are retained in a GEE regression rather than dropped, ensuring that the estimated gaps reflect the population rather than a mobility-selected subsample.

Finally, the GEE method is distribution-agnostic beyond the first two moments. It requires a correct mean function and a variance specification. This quasi-likelihood framework enables population researchers to handle both continuous outcomes, for example, logged income and binary ones, for example, upward-mobility status within the same conceptual framework, and they are implemented in SAS using PROC GENMOD by toggling the DIST and LINK options. In short, GEE aligns the mathematics of estimation with the substantive focus of social mobility research, population-average effects that inform aggregate policy calculations, such as forecasting how much an education grant could narrow income disparities across birth backgrounds. GEE's estimating-equation machinery gives us precisely what policy analysts need: consistent, population-average coefficients that translate directly into statements such as "subjects from the bottom parental-income quintile earn, on average, $X less per year than those from the top quintile, holding age, education and labour-market status constant, and obtaining a Bachelor's degree increases the probability of upward mobility by Y percentage points".

Chapter 3 introduced a synthetically created dataset mimicking the approximate distributional properties of the LAD that will now serve as the empirical engine for our GEE applications in this chapter. This dataset contains 22,660 person-year observations on 5,000 subjects followed for up to five waves. Because simulated migration and employment exit remove some respondents after their departure year, the panel is unbalanced, an ideal proving ground for GEE, which retains subjects even when a few annual records are missing. Table 7.1 lists all variables that are considered for the Gaussian and logit GEE models of this chapter.

Several features of this dataset reinforce the choice of a population-averaged GEE approach. First, the time-invariant parental variables, such as parental

TABLE 7.1

Case Study 3 Overview

Variable	Scale/Levels	Time-Varying?	Intended Role in Models
ID	1 … 5,000	No	Repeated-subject key in REPEATED SUBJECT=clause
Year	2018–2022	Yes	Calendar trend; lags for AR(1) diagnostics
Income	Annual dollars	Yes	Gaussian outcome
Parental income quintile	1 (lowest) … 5 (highest)	No	Core time-invariant mobility predictor
Parental education level	High School, College, University	No	Additional socioeconomic control
Gender	Male, Female	No	Demographic covariate retained under GEE
Age	18–69 (updated annually)	Yes	Life-cycle control (centred at 40 in analysis)
Education level	HighSchool, Tertiary (up-skilling allowed)	Yes	Main policy lever: will interact with the Year
Employment status	Full-time, Part-time, Unemployed	Yes	Labour-market control
Occupation type	Administrative, Managerial, Technical	Yes	Skill-mix control
Industry sector	Services, Public, Manufacturing	Yes	Sectoral control
Region	Ten Canadian provinces plus the Rest of Canada	No	Spatial fixed effect
Household size	1–8 persons	Yes	Household-context control
Migration status	0/1 flag for inter-provincial move	Yes	Attrition predictor; enters weighting check
Economic shocks indicator	1 in 2020 recession, 0 otherwise	Yes	Macro-shock dummy; tests slope stability
Upward mobility (derived)	1 if current quintile > parental quintile	Yes	Secondary logit outcome

income quintile and education level, remain in the design matrix, illustrating a clear advantage over FE models, which would eliminate such predictors through differencing. Second, we will derive a binary "upward mobility" indicator later and analyse it with a logit GEE, complementing the continuous log-income model and giving a second lens on income mobility analysis. Third, because about 8 per cent of subjects provide fewer than five annual observations, the ability of GEE to incorporate every available record and to supply robust standard errors that remain valid under this monotone attrition pattern is essential.

With the dataset's structure and variables clarified, the next step is to run our panel diagnostic checklist, balance counts, and within- and between-variance ratios before specifying the population-average mean models.

This section outlines a tightly focused modelling agenda in which we fit two population-averaged GEE specifications to the Case Study 3 dataset to quantify income and income mobility, and to determine which working-correlation structure best captures the longitudinal dependencies prevalent in the data. Our first GEE specification treats annual income as a continuous outcome, which is estimated with a Gaussian-identity link, while the second specification recodes a binary indicator of upward mobility and analyses that endpoint with a logit link.

For each model, we run the GEE twice, once assuming an exchangeable correlation matrix and once under a first-order autoregressive structure. We run each GEE twice, once with an exchangeable correlation and once with an AR(1) correlation, because the accurate within-subject correlation is unknown. GEE coefficients are robust, but precision depends on this working correlation. Fitting both allows us to select the more efficient option and verify that conclusions are not sensitive to the correlation assumption. Let's dive right into our GEE model development with SAS Syntax 7.1

SAS Syntax 7.1

```
proc freq data=work.Case_Study_3 nlevels;
   tables ID*Year / list nocum nopercent;
   title "Panel balance: person-year counts (WORK.IMPORT)";
run;
proc sql;
   create table WaveCount as
   select ID, count(Year) as Waves
   from  work.Case_Study_3
   group  by ID;
quit;
proc freq data=WaveCount;
```

```
   tables Waves;
   title "Distribution of completed waves per individual";
run;
proc summary data=work.Case_Study_3 (where=(not missing(Income))) nway;
   class ID;
   var  Income;
   output out=WB(drop=_type_ _freq_)
         mean=Mean_Inc std=SD_Inc n=Obs;
run;
proc means data=work.Case_Study_3(where=(not missing(Income))) noprint;
   var Income;
   output out=VarTot(drop=_type_ _freq_) var=Var_Total;
run;
data VarReport;
   if _n_=1 then set VarTot;        /* pull overall variance */
   set WB end=last;
   retain SumWithin Var_Total;
   SumWithin + SD_Inc**2;           /* accumulate within-ID var */
   if last then do;
      Var_Within  = SumWithin / _n_;
      Var_Between = Var_Total  - Var_Within;
      Ratio_BW   = Var_Between / Var_Within;
      keep Var_Total Var_Within Var_Between Ratio_BW;
      output;
   end;
run;
proc print data=VarReport label;
   label Var_Total  = "Overall variance (Income)"
        Var_Within = "Average within-subject variance"
        Var_Between= "Between-subject variance"
        Ratio_BW   = "Between : Within variance ratio";
   title "Variance decomposition for Income";
run;
```

What the Code Does

The code-block begins by checking how well balanced the data is. PROC FREQ cross-tabulates every *ID*Year* combination so readers can see exactly which subjects are present in which survey waves, while a short SQL query counts the number of waves per subject, and a second PROC FREQ prints the distribution of those counts, showing, for example, how many subjects have all five waves versus three, two, or one. In the second part of the code-block, how much income variation comes from differences within a subject over time versus differences between subjects is performed. PROC SUMMARY calculates each subject's mean and standard deviation of income (after omitting

TABLE 7.2

Variance Decomposition for Income

Obs	Overall Variance (Income)	Average Within-Subject Variance	Between-Subject Variance	Between: Within Variance Ratio
1	321,994,110.53	16,629,899.32	305,364,211.2	18.3624

missing values); PROC MEANS captures the total variance in the panel data. A DATA step then adds up all the within-subject variances, converts that sum to an average (*Var_Within*), and subtracts it from the total variance to obtain the between-subject component (*Var_Between*). Finally, the code-block prints a one-row report that lists the overall variance, the average within-subject variance, the between-subject variance, and their ratio. A large ratio signals that most variability lies between subjects, implying strong intra-subject correlation and justifying the use of a working-correlation structure such as exchangeable or AR(1) in subsequent GEE model specifications.

Table 7.2 displays that all of the variance in income arises from cross-sectional differences between subjects rather than from year-to-year changes within subjects. An 18.3624-to-1 ratio means stable differences, for example, parental background, education, region, gender, and long-run occupational track, dominate the variations in income, whilst genuine mobility (year-on-year gains or losses for the same subject) contributes only a modest slice of variability.

Because the focus of the estimation target we employ in this chapter is the population-average mean, GEE estimation will capture those between-subject gaps without being hampered by the limited within-subject variance, which also implies that any reasonable working correlation will have limited influence on point estimates.

SAS Syntax 7.2

```
data work.cs3_prep;
   set work.Case_Study_3;
   Age_c = Age - 40;
   ParIncQ = input(compress(ParentalIncomeQuintile, , 'kd'), best.);
run;
proc sort data=work.cs3_prep; by Year; run;
proc rank data=work.cs3_prep out=work.cs3_rank ties=low groups=5;
   by Year;
   var Income;
   ranks IncQ0;
run;
```

```
data work.cs3_final;
  set work.cs3_rank;
  /*Child's current quintile, numeric 1–5 and labelled "Q1"–"Q5"*/
  ChildIncQ = IncQ0 + 1;
  length CurrentIncQuintile $2;
  CurrentIncQuintile = cats('Q', put(ChildIncQ, 1.));
  /*Upward mobility: 1 if child exceeds parental quintile*/
  Upward_Mobility = (ChildIncQ > ParIncQ);
  /*Log of income (optional, for Model A)*/
  if not missing(Income) then LogInc = log(Income);
run;
```

What the Code Does

The programme creates an analysis-ready longitudinal file in three orderly steps. First, it reads the Case Study 3 dataset and constructs *cs3_prep*: age is centred at 40 (*Age_c*), and the parental income quintile stored as the text "Q1'–'Q5" is transformed to a numeric 1–5 by stripping out everything except digits (COMPRESS(...,'kd')). Second, it sorts the data by survey year and uses PROC RANK with groups=5 to place each child's current income into a year-specific quintile. RANK returns 0–4 in the variable IncQ0; adding 1 yields ChildIncQ=1–5, and ties are assigned to the lower quintile (ties=low), so every year has five ordered groups. Third, the code builds cs3_final by (i) turning *ChildIncQ* into the label "Q1'–'Q5" (*CurrentIncQuintile*), (ii) setting *Upward_Mobility*=1 whenever the child's quintile exceeds the parent's numeric quintile, otherwise (iii) and taking the natural log of income (*LogInc*) for later Gaussian models. The resulting *work.cs3_final* thus contains, for every person-year record, a centred age, numeric and labelled income quintiles for parent and child, a 0/1 upward-mobility indicator, and an optional log-income variable, ready for use in Model A (log-income) and Model B (upward-mobility) analyses.

SAS Syntax 7.3 below will establish the backbone of our GEE model specification for the outcome variable income transformed into the log-income scale. The syntax's generated SAS outputs will be this chapter's first set of analysis outputs.

SAS Syntax 7.3

```
%macro run_gauss(type);
  %let suffix = %sysfunc(compress(&type,'()'));
    proc genmod data = work.cs3_final;
    class ID Year Region EmploymentStatus EducationLevel
        ParentalEducationLevel ParentalIncomeQuintile;
```

```
      model LogInc =
          ParentalIncomeQuintile
          EducationLevel
          EmploymentStatus
          Age_c
          Year
          EconomicShocksIndicator
          Region
          EducationLevel*Year
          / dist = normal link = identity;
      repeated subject = ID / type = &type corrw;
      ods output GEEFitCriteria = Fit_LogInc_&suffix;
      title "Gaussian GEE for log(Income) – working corr = &type";
    run;
  %mend;
  %run_gauss(EXCH);
  %run_gauss(AR(1));
  proc transpose data = Fit_LogInc_EXCH
            out  = FC_EXCH(drop = _NAME_) prefix = EXCH_;
    id  Criterion;
    var Value;
  run;
  proc transpose data = Fit_LogInc_AR1
            out  = FC_AR1(drop = _NAME_)  prefix = AR1_;
    id  Criterion;
    var Value;
  run;
  data QIC_LogInc;
    length WorkCorr $6;
    merge FC_EXCH(in = a) FC_AR1(in = b);
    if a then do WorkCorr = 'EXCH';  QIC = EXCH_QIC;  CIC = EXCH_CIC;  out-
put; end;
    if b then do WorkCorr = 'AR(1)'; QIC = AR1_QIC;   CIC = AR1_CIC;   output;
end;
    keep WorkCorr QIC CIC;
  run;
  proc print data = QIC_LogInc label noobs;
    title "GEE fit criteria for log-income (smaller QIC is better)";
  run;
```

What the Code Does

The code-block defines a reusable macro "%run_gauss" that fits a population-average Gaussian GEE to subjects' log-income (LogInc). The macro takes a single argument specifying the working-correlation structure, either exchangeable (EXCH) or first-order autoregressive (AR(1)). Inside the macro, SAS first sanitises the argument so that AR(1) becomes the suffix AR1. Next, PROC GENMOD models log-income as a linear function of parental income quintile, the subject's education, employment status, centred age, survey year, an economic-shock indicator, region, and an education-by-year interaction. The REPEATED statement clusters observations by subject (*ID*) and applies the requested correlation type, while an ODS OUTPUT directive captures the model-selection statistics (QIC and CIC) in a data set named *Fit_LogInc_<suffix>*. After defining the macro, the programme calls it twice, once with EXCH, once with AR(1), producing two separate fit criteria tables. Each table is then transposed, hence the QIC and CIC values become columns rather than rows, assigned the prefixes EXCH_ or AR1_, and stored in FC_EXCH and FC_AR1. A subsequent DATA step merges the two tables, standardises the variable names, and creates QIC_LogInc, a one-row summary that lists the working-correlation structure alongside its QIC and CIC. Finally, PROC PRINT displays this comparison, enabling the readers to decide which working correlation (exchangeable or AR(1)) offers a more effective quasi-likelihood fit for modelling subjects' income.

Tables 7.3 and 7.4 are the initial results table of the SAS code-block. Table 7.4 describes the GEE Fit Criteria. A QIC of 22,725.7 (QICu=22,688.0) provides a baseline measure of model efficiency where the absolute number is not meaningful on its own, but it becomes decisive once we benchmark it against the QIC from the AR(1) specification. If the AR(1) model yields a lower value, its decaying correlation pattern is more efficient; otherwise, the simpler exchangeable structure is preferred.

Table 7.5 presents the results of the model run, confirming a familiar narrative on income mobility for population researchers. The Gaussian GEE fitted to log-income provides a parsimonious picture of what drives earnings in this cohort. After adjustment for all other covariates and the within-subject

TABLE 7.3

GEE Model Information

Correlation Structure	Exchangeable
Subject effect	ID (5,000 levels)
Number of clusters	5,000
Correlation matrix dimension	5
Max cluster size	5
Min cluster size	1

TABLE 7.4

GEE Fit Criteria

QIC	22725.7445
QICu	22686.0

TABLE 7.5

Analysis of GEE Parameter Estimates: Empirical Standard Error Estimates (Corr Structure = (Exchangeable)

Parameter		Estimate	Standard Error	95% Confidence Limits	Z	Pr>Z
Intercept		10.3038	0.0203	10.2641–10.3436	507.96	<0.0001
Parental Income Quint 5	Q1	–0.0216	0.0192	–0.0593–0.0161	–1.12	0.2617
Parental Income Quint 5	Q2	–0.0263	0.0194	–0.0643–0.0116	–1.36	0.1741
Parental Income Quint 5	Q3	0.0123	0.0198	–0.0265–0.0510	0.62	0.5347
Parental Income Quint 5	Q4	–0.0007	0.0193	–0.0386–0.0372	–0.04	0.9706
Parental Income Quint 5	Q5	0.0000	0.0000	0.0000–0.0000		
Education Level	HighSchool	–0.6010	0.0166	–0.6335–0.5686	–36.30	<0.0001
Education Level	Tertiary	0.0000	0.0000	0.0000–0.0000	.	.
Employment Status	Full-time	0.1240	0.0041	0.1159–0.1321	29.97	<0.0001
Employment Status	Part-time	0.1340	0.0042	0.1258–0.1422	32.07	<0.0001
Employment Status	Unemployed	0.0000	0.0000	0.0000–0.0000	.	.
Age_5		0.0001	0.0005	–0.0010–0.0012	0.17	0.8671
Year	1	0.1021	0.0057	0.0910–0.1132	18.04	<0.0001
Year	2	0.0814	0.0046	0.0724–0.0905	17.62	<0.0001
Year	3	0.0431	0.0035	0.0362–0.0499	12.31	<0.0001
Year	4	0.0201	0.0020	0.0163–0.0239	10.29	<0.0001
Year	5	0.0000	0.0000	0.0000–0.0000		
Economic Shock Index		0.0000	0.0000	0.0000–0.0000		
Region	Alberta	0.0096	0.0181	–0.0259–0.0420	0.36	0.7158
Region	British Columbia	0.0018	0.0189	–0.0352–0.0388	0.10	0.9241

(Continued)

TABLE 7.5 (*Continued*)

Analysis of GEE Parameter Estimates: Empirical Standard Error Estimates (Corr Structure=(Exchangeable)

Parameter		Estimate	Standard Error	95% Confidence Limits	Z	Pr>Z
Region	Manitoba	0.0218	0.0183	−0.0141–0.0577	1.19	0.2337
Region	New Brunswick	0.0265	0.0187	−0.0101–0.0631	1.42	0.1553
Region	Newfoundland and Labrador	−0.0139	0.0179	−0.0490–0.0213	-0.77	0.4398
Region	Nova Scotia	0.0163	0.0182	−0.0193–0.0520	0.90	0.3689
Region	Ontario	0.0113	0.0192	−0.0262–0.0489	0.59	0.5541
Region	Prince Edward Island	−0.0041	0.0181	−0.0396–0.0315	−0.22	0.8226
Region	Quebec	0.0127	0.0168	−0.0202–0.0457	0.76	0.4489
Region	Saskatchewan	0.0000	0.0000	0.0000–0.0000		
Year* Education Level	1 HighSchool	0.0010	0.0104	−0.0193–0.0213	0.10	0.9219
Year* Education Level	1 Tertiary	0.0000	0.0000	0.0000–0.0000		
Year* Education Level	2 HighSchool	0.0006	0.0085	−0.0161–0.0174	0.07	0.9416
Year* Education Level	2 Tertiary	0.0000	0.0000	0.0000–0.0000		
Year* Education Level	3 HighSchool	0.0041	0.0066	−0.0088–0.0170	0.62	0.5338
Year* Education Level	3 Tertiary	0.0000	0.0000	0.0000–0.0000		
Year* Education Level	4 HighSchool	0.0022	0.0037	−0.0051–0.0094	0.58	0.5605
Year* Education Level	4 Tertiary	0.0000	0.0000	0.0000–0.0000		
Year* Education Level	5 HighSchool	0.0000	0.0000	0.0000–0.0000		
Year* Education Level	5 Tertiary	0.0000	0.0000	0.0000–0.0000		

correlation of repeated measures, schooling emerges as the dominant determinant of income. Subjects whose highest qualification is high school earn, on average, 0.601 log-points less than their tertiary-educated peers ($Z=-36.3$, $p<0.0001$). The interaction terms between education and each follow-up year are all small and non-significant, indicating that this schooling gap neither widens nor narrows over the 5-year panel. Labour-force attachment is also found to matter. Relative to being unemployed, holding a full-time position adds 0.124 log-points (approximately 13 per cent), while even part-time work confers a slightly larger premium of 0.134 log-points (approximately 14 per cent) and both effects are statistically significant.

By contrast, parental income quintile, often a strong proxy for socio-economic background, shows no independent association with subjects' earnings once their education and employment status are controlled. Furthermore, the time pattern in wages exhibits a monotonic but attenuating trend, where, compared with the baseline year, the mean log-income rises by 0.102 log points in Year 1, 0.081 in Year 2, 0.043 in Year 3, and 0.020 in Year 4, with no further gain in Year 5. All four non-zero contrasts are found to be statistically significant, indicating an early post-baseline surge followed by a plateau in growth.

Finally, neither age nor the macro-economic-shock index exhibits any detectable effect, and the set of provincial dummies is uniformly non-significant, implying that regional wage differentials disappear once subject-level features are taken into account.

These high-level results suggest that the return to human capital accumulation (education) and current employment status outweigh the influence of factors such as family background, age, region, or short-term macro-economic shocks. Moreover, the education premium remains remarkably time-invariant over time, underscoring the persistent role of attaining formal qualifications in structuring labour-market returns.

Similar to our previous model specification with exchangeable working correlation, our Gaussian GEE specification with first-order autoregressive working-correlation structure confirms that the GEE analysis is being estimated on the complete 22,660 person-year observations. Treating the outcome as *LogInc* and pairing a Gaussian distribution with the identity link informs the reader that the model targets the population-average change in the natural log of income for a one-unit shift in each predictor, for example, exponentiating any coefficient therefore yields an approximate percentage effect on gross income.

The GEE Fit Criteria table output for the GEE model with the Corr Structure=AR (1) displays a QIC=22,703.7, along with its small-sample analogue, QICu=22,686. Because they are lower than the 22,726/22,688 recorded under the exchangeable model, the AR(1) specification provides a more efficient population-average fit for log-income, as it better captures the way serial dependence decays with time while leaving parameter estimates unbiased. In addition, the gap between QIC and

QICu indicates that the panel dataset has an effective sample size for small-sample corrections to be negligible.

Under the more plausible AR(1) working correlation, the Gaussian GEE reproduces the substantive story obtained with the exchangeable structure. Still, it yields slightly tighter empirical standard errors, confirming that efficiency improves when the correlation pattern mirrors the data's time ordering. Point estimates, as shown in Table 7.6, are essentially unchanged from the previous model with Corr Structure = (Exchangeable). The model intercept is 10.309 ± 0.017 log-points, and the education coefficient for leaving school remains large. On the original scale, this corresponds to e^−0.601 ≈ 0.55, that is, a 45 per cent wage penalty for high school leavers relative to the tertiary-educated, a gap that remains time-invariant over the years, mainly because every *Year * Education* interaction is not statistically significant.

The labour-force status retains its dominant role, with full-time employment adding 0.122 log-points, and part-time work conferring a virtually identical 0.127 log-points, underscoring the precision gained with AR(1). The parental income quintile still shows no independent association (all $|Z| \leq 1.32$), demonstrating that, once own human capital and job status are controlled, family background offers no additional predictive power for earnings. The cohort's temporal profile is also robust, with wages jumping by 10–11 per cent in Year 1, slipping to an 8 per cent gain in Year 2, and then tapering to 4 per cent (Year 3) and 2 per cent (Year 4) before flattening, with every non-zero contrast highly significant ($|Z| \geq 11$). Age, the economic-shock indicator, and all but one provincial dummy remain negligible, with only New Brunswick recording a small premium of 0.021 log-points, barely crossing the 5 per cent threshold ($Z = 2.07$, $p = 0.039$).

In short, what we observe is the impact of imposing an AR(1) correlation structure. The evidence continues to indicate that formal qualifications and attachment to the labour market govern earnings trajectories, and that the sizeable education premium remains constant.

Upward mobility is a binary outcome that can be observed repeatedly for the same subject across multiple years. Because many subjects never switch status (they are either always above or always below their parents' income quintile), including them would create empty or quasi-empty cells and make the Hessian matrix singular. By restricting the analysis to movers, for example, subjects who exhibit at least one 0 and one 1, in this section, we will guarantee genuine within-cluster variation, allowing our GEE model specifications to estimate a full covariance matrix and compare correlation structures (independent, exchangeable, AR(1)) via the QIC. The result is expected to be a robust, interpretable model of the drivers of upward mobility that reports policy-oriented, population-average effects. SAS Syntax 7.4 below frames the SAS code-block of such a methodical approach.

TABLE 7.6

Analysis of GEE Parameter Estimates: Empirical Standard Error Estimates (Corr Structure=(AR(1))

Parameter		Estimate	Standard Error	95% Confidence Limits	Z	Pr>Z
Intercept		10.3093	0.0165	10.2770–10.3416	624.53	<0.0001
Parental Income Quint 5	Q1	–0.0215	0.0193	–0.0594–0.0164	–1.11	0.2661
Parental Income Quint 5	Q2	–0.0256	0.0194	–0.0636–0.0124	–1.32	0.1861
Parental Income Quint 5	Q3	0.0132	0.0198	–0.0256–0.0519	0.67	0.5047
Parental Income Quint 5	Q4	0.0006	0.0193	–0.0373–0.0384	0.03	0.9758
Parental Income Quint 5	Q5	0.0000	0.0000	0.0000–0.0000		
EducationLevel	HighSchool	–0.6013	0.0164	–0.6335–0.5691	–36.58	<0.0001
Education Level	Tertiary	0.0000	0.0000	0.0000–0.0000		
Employment Status	Full-time	0.1221	0.0017	0.1188–0.1253	73.48	<0.0001
Employment Status	Part-time	0.1267	0.0018	0.1235–0.1299	76.89	<0.0001
Employment Status	Unemployed	0.0000	0.0000	0.0000–0.0000		
Age_5		0.0001	0.0005	–0.0010–0.0012	0.20	0.8444
Year	1	0.1025	0.0056	0.0916–0.1134	18.44	<0.0001
Year	2	0.0818	0.0045	0.0730–0.0906	18.20	<0.0001
Year	3	0.0434	0.0033	0.0369–0.0500	13.01	<0.0001
Year	4	0.0200	0.0018	0.0164–0.0235	11.03	<0.0001
Year	5	0.0000	0.0000	0.0000–0.0000		
Economic Shock Index		0.0000	0.0000	0.0000–0.0000		
Region	Alberta	0.0068	0.0093	–0.0113–0.0250	0.74	0.4617
Region	British Columbia	–0.0043	0.0098	–0.0234–0.0149	–0.43	0.6638

(Continued)

TABLE 7.6 (*Continued*)

Analysis of GEE Parameter Estimates: Empirical Standard Error Estimates (Corr Structure=(AR(1))

Parameter		Estimate	Standard Error	95% Confidence Limits	Z	Pr>Z
Region	Manitoba	0.0084	0.0096	–0.0104–0.0272	0.88	0.3794
Region	New Brunswick	0.0209	0.0101	0.0011–0.0407	2.07	0.0387
Region	Newfoundland and Labrador	–0.0036	0.0095	–0.0221–0.0150	–0.38	0.7045
Region	Nova Scotia	0.0096	0.0095	–0.0090–0.0281	1.01	0.3131
Region	Ontario	0.0057	0.0095	–0.0129–0.0242	0.60	0.5494
Region	Prince Edward Island	–0.0054	0.0099	–0.0248–0.0139	–0.55	0.5824
Region	Quebec	0.0092	0.0091	–0.0085–0.0270	1.02	0.3085
Region	Saskatchewan	0.0000	0.0000	0.0000–0.0000		
Year* Education Level	1 HighSchool	0.0011	0.0102	–0.0189–0.0210	0.11	0.9152
Year* Education Level	1 Tertiary	0.0000	0.0000	0.0000–0.0000		
Year* Education Level	2 HighSchool	0.0008	0.0083	–0.0155–0.0171	0.10	0.9224
Year* Education Level	2 Tertiary	0.0000	0.0000	0.0000–0.0000		
Year* Education Level	3 HighSchool	0.0040	0.0063	–0.0084–0.0164	0.63	0.5292
Year* Education Level	3 Tertiary	0.0000	0.0000	0.0000–0.0000		
Year* Education Level	4 HighSchool	0.0017	0.0034	–0.0050–0.0085	0.50	0.6148
Year* Education Level	4 Tertiary	0.0000	0.0000	0.0000–0.0000		
Year* Education Level	5 HighSchool	0.0000	0.0000	0.0000–0.0000		
Year* Education Level	5 Tertiary	0.0000	0.0000	0.0000–0.0000		

SAS Syntax 7.4

```
data work.B_prep;
   set work.cs3_final;
   /* alias for clarity */
   UpMob = Upward_Mobility;
   /* numeric employment flag */
   if EmploymentStatus in ('Full-time','Part-time') then Employment_bin = 1;
   else if EmploymentStatus = 'Unemployed'        then Employment_bin = 0;
   /* collapse small provinces */
   length Region_collapsed $17;
   if Region in ('Ontario','Quebec','British Columbia')
       then Region_collapsed = Region;
   else Region_collapsed = 'Other';
   if cmiss(EducationLevel, Region_collapsed)=0
     and nmiss(UpMob, Employment_bin, Age_c, Year)=0;
run;
proc sql noprint;
   create table movers_id as
   select ID
   from   work.B_prep
   group  by ID
   having min(UpMob)=0 and max(UpMob)=1;
quit;
proc sort data=work.B_prep ; by ID ; run;
data work.gee_sample;
   merge work.B_prep (in=data)
        movers_id   (in=keep);
   by ID;
   if keep;
run;
%macro run_gee(type);
   %let suff = %sysfunc(compress(&type,'()'));
   proc genmod data=work.gee_sample descending;
      class ID Year Region_collapsed EducationLevel ParentalIncomeQuintile;
      model UpMob = ParentalIncomeQuintile
                  EducationLevel
                  Employment_bin
                  Age_c
                  Year
                  Region_collapsed
                  / dist=binomial link=logit type3;
      repeated subject=ID / type=&type;
      ods output GEEFitCriteria = FitMob_&suff;
```

```
    title "GEE logit for Upward Mobility – working corr =&type";
  run;
%mend;
%run_gee(IND);
%run_gee(EXCH);
%run_gee(AR(1));
data QIC_Mobility;
  merge FitMob_IND  (where=(Criterion='QIC') rename=(Value=QIC_IND))
      FitMob_EXCH (where=(Criterion='QIC') rename=(Value=QIC_EXCH))
      FitMob_AR1  (where=(Criterion='QIC') rename=(Value=QIC_AR1));
run;
proc print data=QIC_Mobility label noobs;
  label QIC_IND='IND' QIC_EXCH='EXCH' QIC_AR1='AR(1)';
  title 'Model B – QIC by working correlation (smaller = better)';
run;
```

What the Code Does

The code-block first turns the cleaned longitudinal file *cs3_final* into an analysis-ready panel "B_prep" by first recoding employment into a numeric 0/1 flag, then collapsing sparsely represented provinces into a single "Other" category, later centring age, and finally discarding rows with missing values so that every remaining record contains a complete set of predictors. The code-block then isolates movers, which are subjects whose upward-mobility flag switches at least once, by asking, for each ID, whether the minimum is 0 and the maximum is 1, where every wave for those mover-IDs is merged back to form *gee_sample*, ensuring each selected subject contributes a full block of repeated measures. The macro, "%run_gee", is called three times throughout the run-time of the code-block, once for each working-correlation structure (independent, exchangeable, AR(1)). In each call PROC GENMOD fits a binomial GEE that models the probability of upward mobility as a function of parental income quintile, the subject's education, employment status, centred age, survey year, and collapsed region, clustering on ID and exporting the QIC and CIC to a fit table whose name reflects the correlation type. Finally, three fit tables are merged into "QIC_Mobility"" and the programme prints a single row showing the QIC for IND, EXCH, and AR(1), so the readers can see which working correlation yields the most efficient, population-average estimates for our second GEE model.

Tables 7.7 and 7.8 summarise the model run under the assumption of independent observations across repeated measures. While independence is rarely true in longitudinal data, this structure ensures consistent point estimates and can be used as a baseline for efficiency comparisons. The QIC is 3819.2, and QICu is 3817.8, values that serve as benchmarks against which to evaluate the relative efficiency of the exchangeable and AR(1) structures.

TABLE 7.7

Model Information (Corr Structure=IND)

Parameter	Value
Correlation structure	Independent
Subject effect	ID (585 levels)
Number of clusters	585
Correlation matrix dimension	5
Maximum cluster size	5
Minimum cluster size	2

TABLE 7.8

GEE Fit Criteria (Corr Structure=IND)

Criterion	Value
QIC	3819.1788
QICu	3817.8291

Lower values indicate better quasi-likelihood fit, conditional on the same model structure and covariates.

The results from the GEE logistic regression model, as presented on Table 7.9, indicate several key predictors of upward income mobility, defined as achieving a higher income quintile than one's parents. The intercept is significantly negative (–0.7225, p <.0001), implying that the baseline subject, for example, someone from parental income quintile 4 with tertiary education, unemployed, aged 40, residing in Quebec, and observed in year 5 has below-average odds of upward mobility. Parental income background appears to exert a significant influence: individuals from Q1 have substantially higher odds of upward mobility compared to the reference group (Q4), with a log-odds increase of 0.41 (p=0.0014); those from Q2 also benefit. However, differences between Q3 and Q4 are not statistically significant.

Education emerges as a strong determinant – those with only a high school qualification are significantly less likely to experience upward mobility compared to their peers with a tertiary education. Employment status is the most potent positive predictor in the model: being employed is associated with a near unit increase in log-odds, indicating a substantial increase in the probability of mobility. Age, centred at 40, does not have a significant effect. Calendar-year indicators reveal no significant time trend in mobility after adjusting for other factors.

Lastly, regional effects are negligible, with none of the provinces (British Columbia, Ontario, or Other regions) differing significantly from Quebec. Taken together, the model highlights the significance of socioeconomic background, educational attainment, and labour-market participation in shaping individuals' opportunities for upward economic mobility.

TABLE 7.9

Analysis of GEE Parameter Estimates: Empirical Standard Error Estimates (Corr Structure=IND)

Parameter	Estimate	Standard Error	95% Confidence Limits	Z	Pr> |Z|
Intercept	−0.7225	0.1800	−1.0753–0.3698	−4.01	<0.0001
Parental Income Quint 5 Q1	0.4099	0.1285	0.1579–0.6618	3.19	0.0014
Parental Income Quint 5 Q2	0.2724	0.1182	0.0407–0.5041	2.30	0.0212
Parental Income Quint 5 Q3	0.0209	0.1206	−0.2155–0.2573	0.17	0.8626
Parental Income Quint 5 Q4	0.0000	0.0000	0.0000–0.0000		
Education Level High School	−0.4272	0.1069	−0.6368–0.2176	−4.00	<0.0001
Education Level Tertiary	0.0000	0.0000	0.0000–0.0000		
Employment_bin	0.9731	0.0907	0.7953–1.1508	10.73	<0.0001
Age_c	0.0013	0.0036	−0.0058–0.0084	0.36	0.7218
Year 1	−0.2415	0.1617	−0.5583–0.0754	−1.49	0.1353
Year 2	−0.0891	0.1410	−0.3655–0.1873	−0.63	0.5276
Year 3	0.0080	0.1233	−0.2338–0.2497	0.06	0.9483
Year 4	0.0123	0.0905	−0.1651–0.1896	0.14	0.8920
Year 5	0.0000	0.0000	0.0000–0.0000		
Region_collapsed British Columbia	−0.1940	0.1761	−0.5392–0.1511	−1.10	0.2705
Region_collapsed Ontario	−0.0818	0.1947	−0.4634–0.2998	−0.42	0.6745
Region_collapsed Other	0.0225	0.1394	−0.2508–0.2957	0.16	0.8721
Region_collapsed Quebec	0.0000	0.0000	0.0000–0.0000		

Table 7.10 reports Type III score statistics (Wald Chi-Square tests) from a GEE model assessing the joint significance of predictor variables in explaining upward income mobility. The results confirm that Parental Income Quintile, Education Level, and Employment status are all statistically significant contributors to upward mobility at conventional thresholds.

To assess whether a more efficient model fit could be achieved, we now turn to the GEE specification using an exchangeable working-correlation structure, which assumes that all within-subject correlations are equal regardless of the time lag between observations.

Table 7.11 presents exchangeable working condition of our model whilst Table 7.12 shows model's fit criteria. According to this specification, the estimated average within-subject correlation is extremely low (0.0082), indicating

TABLE 7.10

Analysis of GEE Parameter Estimates: Score Statistics for Type 3 (Corr Structure = IND)

Source	DF	Chi-Square	Pr>ChiSq
Parental Income Quint 5	3	15.53	0.0014
Education Level	1	15.67	<0.0001
Employment_bin	1	107.36	<0.0001
Age_c	1	0.13	0.7220
Year	4	3.87	0.4246
Region_collapsed	3	3.05	0.3837

TABLE 7.11

Exchangeable Working Correlation (Corr Structure = EXCH)

Parameter	Value
Correlation	0.0081669401

TABLE 7.12

GEE Fit Criteria (Corr Structure = EXCH)

Criterion	Value
QIC	3819.2832
QICu	3817.8462

that the repeated measures of upward-mobility status across years are nearly independent after accounting for covariates. This minimal correlation undermines the advantage typically gained from assuming exchangeability, as it indicates little residual dependence between repeated outcomes. The model's QIC is 3,819.28, with a corrected QICu of 3,817.85. These values are virtually identical to those from the independence model, implying no efficiency gain and no discernible improvement in model fit when assuming exchangeability. Thus, the data provides little justification for adopting this more complex correlation structure over the simpler assumption of independence.

Table 7.13 presents parameter estimates from a GEE logistic regression model using the exchangeable working-correlation structure. The intercept is significantly negative (–0.7317, $p < 0.0001$), indicating that the baseline individual, who is from parental income quintile Q4, has tertiary education, is unemployed, is at the centred age of 40, is in Year 5, and resides in Quebec, has a below-average probability of upward mobility. Compared to this reference, individuals from the lowest parental income quintile exhibit a significant increase in the log-odds of upward mobility, as do those from Q2, indicating that a lower parental economic background is associated with a

TABLE 7.13

Analysis of GEE Parameter Estimates: Empirical Standard Error Estimates (Corr Structure=EXCH)

Parameter	Estimate	Standard Error	95% Confidence Limits	Z	Pr> |Z|
Intercept	–0.7317	0.1803	–1.0850–0.3784	–4.06	<0.0001
Parental Income Quint 5 Q1	0.4116	0.1288	0.1593–0.6640	3.20	0.0014
Parental Income Quint 5 Q2	0.2733	0.1184	0.0413–0.5054	2.31	0.0210
Parental Income Quint 5 Q3	0.0209	0.1208	–0.2158–0.2577	0.17	0.8624
Parental Income Quint 5 Q4	0.0000	0.0000	0.0000–0.0000		
Education Level High School	–0.4297	0.1071	–0.6396–0.2198	–4.01	<0.0001
Education Level Tertiary	0.0000	0.0000	0.0000–0.0000		
Employment_bin	0.9844	0.0910	0.8060–1.1628	10.82	<0.0001
Age_c	0.0013	0.0036	–0.0059–0.0084	0.35	0.7244
Year 1	–0.2426	0.1619	–0.5600–0.0747	–1.50	0.1340
Year 2	–0.0897	0.1412	–0.3664–0.1871	–0.64	0.5254
Year 3	0.0084	0.1235	–0.2337–0.2505	0.07	0.9459
Year 4	0.0131	0.0906	–0.1646–0.1907	0.14	0.8855
Year 5	0.0000	0.0000	0.0000–0.0000		
Region_collapsed British Columbia	–0.1923	0.1767	–0.5387–0.1541	–1.09	0.2766
Region_collapsed Ontario	–0.0809	0.1953	–0.4636–0.3018	–0.41	0.6787
Region_collapsed Other	0.0235	0.1399	–0.2506–0.2976	0.17	0.8664
Region_collapsed Quebec	0.0000	0.0000	0.0000–0.0000		

greater scope for upward movement. Quintile Q3, however, does not differ significantly from Q4, suggesting that most intergenerational mobility contrast is concentrated at the bottom end.

Educational attainment remains a major determinant: those with only a high school qualification have significantly lower odds of moving upward compared to their peers who have a tertiary education. Likewise, employment status continues to exhibit the strongest positive association, indicating that participation in the labour market is a critical enabler of upward movement. In contrast, age (centred) has no significant effect, and year dummies (Years 1–4 relative to Year 5) do not yield significant differences, implying that the probability of mobility is temporally stable over the observed panel.

TABLE 7.14

Analysis of GEE Parameter Estimates: Score Statistics for Type 3 (Corr Structure=EXCH)

Source	DF	Chi-Square	Pr>ChiSq
ParentalIncomeQuint5	3	15.67	0.0013
EducationLevel	1	15.90	<0.0001
Employment_bin	1	106.21	<0.0001
Age_c	1	0.12	0.7246
Year	4	3.90	0.4198
Region_collapsed	3	3.01	0.3904

Finally, we observe via the results shown on Table 7.14 that regional effects are not statistically significant compared to Quebec. This suggests that, after adjusting for individual-level characteristics, geography does not substantially shape upward-mobility prospects in this model. Overall, the findings under the exchangeable correlation structure confirm the dominant influence of individual education and employment over structural or time-varying factors.

The joint test for parental income quintile is statistically significant, indicating that parental income background plays a meaningful role in shaping upward mobility. Education Level also yields a highly significant result, reinforcing the interpretation that post-secondary education is a key enabler of mobility. The Employment variable stands out as the most influential factor, underscoring the central role of labour-market attachment in fostering economic progression. In contrast, Age, Year, and Region do not significantly contribute to model fit. This suggests that, after accounting for individual socioeconomic characteristics, neither life stage, panel year, nor region of residence has a statistically significant influence on the probability of upward mobility.

We now turn to a third order correlation structure via the first order autoregressive (AR(1)), by examining Tables 7.15 and 7.16, which assumes that within-subject correlations decline exponentially as the time lag between repeated measurements increases. This structure is especially suitable for longitudinal data, where temporally adjacent observations are more similar to each other than those further apart. Under the AR(1) specification, the model comprises 585 clusters (subjects), with between 2 and 5 observations per subject. The working-correlation matrix reflects a structure of dimension 5, corresponding to the panel length. The fit statistics for this model reveal a QIC of 3825.84 and a QICu of 3827.23. These values are higher than those observed under the independence and exchangeable structures, suggesting that the AR(1) assumption does not lead to a more efficient or better-fitting model in this context. Despite its theoretical appeal for time-structured data, the empirical gain from adopting the AR(1) structure here appears negligible, and may in fact reduce parsimony without improving explanatory power.

TABLE 7.15

Model Information (Corr Structure=AR(1))

Parameter	Value
Correlation Structure	AR(1)
Subject Effect	ID (585 levels)
Number of Clusters	585
Correlation Matrix Dimension	5
Maximum Cluster Size	5
Minimum Cluster Size	2

TABLE 7.16

GEE Fit Criteria (Corr Structure=AR(1))

Criterion	Value
QIC	3825.8434
QICu	3827.2267

Table 7.17 describes that, under the AR(1) structure, the pattern of results remains highly consistent: education and employment status are the most salient drivers, while age, year, and region do not exert a detectable influence. The slightly more substantial employment effect observed reflects AR(1) model's alignment with the temporal dynamics in the data.

The intercept is negative, indicating that the baseline subject, for example, someone in parental income quintile 4, with a tertiary education, unemployed, age centred at 40, in Year 5, and residing in Quebec, has below-average odds of upward mobility. Compared to this reference group, subjects from Q1 show a positive association with upward mobility, as do those from Q2. In contrast, Q3 is not significantly different from Q4, echoing previous findings that most contrast in mobility is observed at the lower end of the parental income distribution.

Education again emerges as a robust predictor. Individuals with only a high school education experience significantly lower odds of upward mobility compared to those with a tertiary education. The effect of employment is notably large and highly significant, reinforcing that labour-force participation substantially enhances the likelihood of moving up the income ladder. Age, centred at 40, shows no statistically significant effect ($p=0.6784$), and all year dummies are likewise insignificant, suggesting temporal stability in mobility probabilities after accounting for covariates. Finally, no regional variable differs significantly from Quebec, suggesting limited geographic variation in upward mobility once individual-level characteristics are controlled.

The Type III score statistics, as presented on Table 7.18, from the GEE model with AR(1) correlation structure show that parental income quintile

TABLE 7.17

Analysis of GEE Parameter Estimates: Empirical Standard Error Estimates (Corr Structure = AR(1))

Parameter	Estimate	Standard Error	95% Confidence Limits	Z	Pr> |Z|
Intercept	–0.9055	0.1762	–1.2508–0.5601	–5.14	<0.0001
Parental Income Quint 5 Q1	0.4121	0.1172	0.1825–0.6417	3.52	0.0004
Parental Income Quint 5 Q2	0.2734	0.1050	0.0675–0.4793	2.60	0.0093
Parental Income Quint 5 Q3	0.0276	0.1048	–0.1778–0.2329	0.26	0.7925
Parental Income Quint 5 Q4	0.0000	0.0000	0.0000–0.0000		
Education Level High School	–0.4455	0.0941	–0.6299–0.2611	–4.73	<0.0001
Education Level Tertiary	0.0000	0.0000	0.0000–0.0000		
Employment_bin	1.2330	0.0894	1.0577–1.4082	13.79	<0.0001
Age_c	0.0013	0.0032	–0.0050–0.0077	0.41	0.6784
Year 1	–0.2866	0.1661	–0.6121–0.0389	–1.73	0.0844
Year 2	–0.1264	0.1441	–0.4088–0.1559	–0.88	0.3801
Year 3	0.0074	0.1258	–0.2391–0.2539	0.06	0.9528
Year 4	0.0188	0.0916	–0.1607–0.1984	0.21	0.8371
Year 5	0.0000	0.0000	0.0000–0.0000		
Region_collapsed British Columbia	–0.1504	0.1762	–0.4957–0.1948	–0.85	0.3932
Region_collapsed Ontario	–0.0887	0.1887	–0.4586–0.2812	–0.47	0.6384
Region_collapsed Other	0.0393	0.1404	–0.2359–0.3146	0.28	0.7793
Region_collapsed Quebec	0.0000	0.0000	0.0000–0.0000		

(p=0.0002), education level (p<0.0001), and employment status (p<0.0001) are all statistically significant predictors of upward income mobility. These results indicate that individuals from lower-income backgrounds, with higher education, and those who are employed are significantly more likely to move up the income distribution. In contrast, age (p=0.6731), year (p=0.2735), and region (p=0.4209) do not significantly influence mobility, suggesting that personal socioeconomic factors are far more important than time, age, or geography in shaping mobility outcomes.

Table 7.19 compares the model fit across different working-correlation structures using the QIC, where lower values indicate better fit. The independence structure (QIC=3819.18) yields the best fit, followed very closely by the exchangeable structure (QIC=3819.28). The AR(1) structure, by contrast,

TABLE 7.18

Analysis of GEE Parameter Estimates: Score Statistics for Type 3 (Corr Structure =AR(1))

Source	DF	Chi-Square	Pr>ChiSq
Parental Income Quint 5	3	19.23	0.0002
Education Level	1	22.37	<0.0001
Employment_bin	1	149.67	<0.0001
Age_c	1	0.18	0.6731
Year	4	5.14	0.2735
Region_collapsed	3	2.82	0.4209

TABLE 7.19

QIC by Working Correlation (Smaller = Better)

Criterion	IND	EXCH	AR(1)
QIC	3,819.1788	3,819.2832	3,825.8434

results in a notably higher QIC (3825.84), indicating a poorer fit relative to the others. These results suggest that the more complex correlation assumptions of EXCH and AR(1) offer no advantage compared to the simpler independence assumption for this dataset.

The empirical findings from this chapter have clear and actionable policy implications for governments seeking to promote intergenerational income mobility. By deploying population-averaged models using GEEs, the analysis explicitly targets the marginal effects of structural socioeconomic variables, both logged income and the probability of upward mobility. The choice of GEE as the estimation framework ensures that these insights generalise to the average population, not just to conditional subgroups, making them particularly relevant for public policy design and evaluation.

First and foremost, this chapter provides robust statistical confirmation that education is the central policy lever for improving both earnings and mobility outcomes. Across both continuous (log-income) and binary (upward-mobility) models, having a tertiary education consistently emerged as the strongest predictor of higher income and increased mobility. The wage premium associated with tertiary education remained stable across all survey years and was robust to changes in the correlation structure, resulting in approximately 45 per cent higher earnings for tertiary graduates compared to high school leavers.

Likewise, the logit GEE models showed that individuals with only a high school qualification were significantly less likely to exceed their parental income quintile. These findings suggest that expanding access to post-secondary education, especially for youth from lower-income families, remains a crucial and empirically justified pathway to promoting upward mobility on a large scale. Second, the models show that labour-market participation is another dominant determinant of economic outcomes. Being employed, even part-time, was strongly associated with both higher earnings and greater upward mobility. Notably, the marginal effect of employment status on the log-odds of upward mobility exceeded that of parental income, education, or region.

These results indicate that, in addition to investing in education, public policy must also prioritise inclusive labour-market access. Policies such as job guarantee programmes, employment subsidies, or vocational bridging schemes that facilitate entry into stable employment could yield outsized gains in promoting mobility, particularly for young adults and marginalised populations. Third, while parental income quintile remains a significant predictor of upward mobility, especially for individuals from the lowest two quintiles, it becomes statistically insignificant in predicting adult income once education and employment status are controlled. This result reveals a nuanced but important message for policymakers: although family background matters for shaping initial conditions, its independent effect on long-run income can be effectively mediated through accessible and equitable educational and labour-market systems. Therefore, policy attention should shift from background-targeted interventions to structural investments in human capital formation and work integration.

Notably, the analysis also demonstrates that macro-level factors, such as year and region, exert little to no influence on income or mobility once individual characteristics are taken into account. This suggests that mobility-enhancing policies can be nationally scaled without necessarily needing fine regional tailoring, although sub-provincial disparities not captured in this model may still warrant targeted approaches. Moreover, the non-significance of temporal effects indicates that, over the observed 5-year panel, there was no systematic drift in the probability of upward mobility. This may reflect a structurally stable but rigid mobility regime, where broad societal conditions (e.g., recession) have limited impact absent individual-level human capital. Finally, the modelling exercises show that the choice of working-correlation structure (independent, exchangeable, AR(1)) has minimal effect on policy conclusions. The QIC values and coefficient stability across correlation assumptions confirm that population-average estimates are robust to assumptions about within-subject dependence. This gives policymakers confidence that the findings are not artefacts of modelling choices, but reflect substantive and persistent patterns in the population.

Taken together, the results of this chapter lead to a clear and evidence-based prescription: public investments in tertiary education and access to employment are the most direct and effective mechanisms for enhancing upward mobility, especially among those from low-income backgrounds. Interventions should prioritise removing financial, institutional, and informational barriers to higher education and ensuring that transitions from education to work are smooth, equitable, and inclusive. While family background continues to influence starting positions, its long-term effects can be attenuated through accessible and sustained investments in people's capabilities and work opportunities. Policymakers seeking to narrow income disparities and foster an economically mobile society should base their strategies on these pillars.

Reflection Questions and Answers

Question 1 – Why is GEE a suitable method for studying intergenerational income mobility in this chapter?
Answer 1 – GEE is appropriate because it models population-average responses rather than individual-specific trajectories. This aligns with the policy-oriented objective of identifying average effects of structural variables, such as education and parental background, on income and mobility outcomes across a cohort. It retains time-invariant covariates and accommodates unbalanced panel data, making it ideal for estimating marginal effects that inform the design of aggregate policies.

Question 2 – What is the difference between the subject-specific models used in Chapters 5–6 and the GEE models used in this chapter?
Answer 2 – Subject-specific models, such as FE and RE models, focus on within-subject changes and remove time-invariant variables. GEE, by contrast, retains these variables and estimates the average impact across the population, which is more appropriate when policy interest lies in average outcomes rather than individual dynamics.

Question 3 – What does the variance decomposition analysis reveal about the structure of income variation in the panel?
Answer 3 – The analysis reveals that income variation is dominated by between-subject differences (e.g., education, parental background, region), with a between-to-within variance ratio of approximately 18.4. This justifies the use of GEE, which effectively captures such cross-sectional heterogeneity at the population level.

Question 4 – Which predictors were found to be most strongly associated with higher income in the Gaussian GEE models?
Answer 4 – Education level and employment status were the most powerful predictors. Specifically, tertiary education is associated with a substantial income premium (~45%), and both full-time and part-time employment increase income significantly relative to unemployment. Parental income quintile had no significant effect on income after controlling for education and employment.

Question 5 – What do the logit GEE results reveal about the predictors of upward income mobility?
Answer 5 – The probability of upward mobility increases significantly with higher education and being employed. Individuals from the lowest parental income quintiles also exhibit a greater likelihood of upward mobility, indicating more opportunities for advancement. However, age, survey year, and regional variables had no significant effect.

Question 6 – How do the results change across different working-correlation structures?
Answer 6 – The point estimates remain remarkably stable across all structures, demonstrating robustness. The QIC comparisons indicate that the independent structure offers the best or equivalent model fit, suggesting minimal within-subject correlation once covariates are considered. Thus, conclusions do not hinge on the choice of correlation structure.

Question 7 – Why is the upward-mobility model restricted to "movers" only?
Answer 7 – Including only individuals who switch mobility status (from 0 to 1 or vice versa) ensures sufficient within-subject variation. This is necessary for estimating the covariance structure reliably and avoiding singular Hessian matrices, which could arise if the model includes clusters with constant outcomes.

Question 8 – How does the education-by-year interaction inform our understanding of returns to education?
Answer 8 – The interaction terms between education and survey year were statistically insignificant, indicating that the income advantage of tertiary education remains stable over time, suggesting that the education premium is neither diminishing nor growing across the observed 5-year panel.

Question 9 – What do the policy implications suggest regarding the role of family background in income outcomes?
Answer 9 – While family background (measured by parental income quintile) influences upward mobility, it does not independently predict income once education and employment are accounted for. This implies that structural investments in human capital can mitigate initial disparities and level the playing field.

Question 10 – Summarise the overall policy message of this chapter in one sentence.
Answer 10 – Investments in higher education and inclusive employment opportunities are the most direct and effective strategies for enhancing intergenerational income mobility, outweighing the long-term influence of family background, geography, or macroeconomic conditions.

Bibliography

Ahn, S. (2024). Bayesian inference of longitudinal binary and ordinal data utilizing multivariate probit models. https://doi.org/10.37099/mtu.dc.etdr/1839

Arnau, J., Balluerka, N., Bono, R., & Gorostiaga, A. (2010). General linear mixed model for analysing longitudinal data in developmental research. *Perceptual and Motor Skills*, 110 (2), 547–566. https://doi.org/10.2466/PMS.110.2.547-566

Balázsi, L., Baltagi, B. H., Mátyás, L., & Pus, D. (2024). Random effects models. https://doi.org/10.1007/978-3-031-49849-7_3

Ballinger, G. A. (2004). Using generalized estimating equations for longitudinal data analysis. *Organizational Research Methods*, 7 (2), 127–150. https://doi.org/10.1177/1094428104263672

Bechtel, G. G. (2021). Panel regression of arbitrarily distributed responses. *Journal of Data Science*. https://doi.org/10.6339/JDS.2009.07(2).459

Biddle, N., Breunig, R., & Markham, F. (2019). Describing the top of the income distribution in Australia (CSRM Working Paper No. 4/2019). Centre for Social Research and Methods, Australian National University. https://polis.cass.anu.edu.au/research/publications/working-papers

Biddle, N., & Marasinghe, D. (2021). Using Census, Social Security and tax data from the Multi-Agency Data Integration Project (MADIP) to impute the complete Australian income distribution. SSRN. https://doi.org/10.2139/ssrn.3826001

Bowles, R. P. (2014). Review of advances in longitudinal methods in the social and behavioral sciences, edited by Jeffrey R. Harring and Gregory R. Hancock. *Structural Equation Modeling*, 21 (4), 651–652. https://doi.org/10.1080/10705511.2014.919830

Butler, D. C., Paige, E., Welsh, J., Law, H. D., Moon, L., Banks, E., & Korda, R. J. (2022). Factors related to under-treatment of secondary cardiovascular risk, including primary healthcare: Australian National Health Survey linked data analysis. *Australian and New Zealand Journal of Public Health*, 46 (4), 533–539. https://doi.org/10.1111/1753-6405.13254

Broadbent, J. M., & Osborne, A. J. (2022). Longitudinal research in Aotearoa New Zealand. *Journal of the Royal Society of New Zealand*. https://doi.org/10.1080/03036758.2022.2090413

Cowie, L., & Hendrickson, L. (2020). Higher education completion rates for domestic undergraduate students who access mental health services using a multi-agency matched population cohort. *International Journal of Population Data Science*, 5 (5). https://doi.org/10.23889/ijpds.v5i5.1502

Demystifying longitudinal data analyses using structural equation models in school psychology. https://doi.org/10.31234/osf.io/pv8bk

Diggle, P. J., Heagerty, P. J., Liang, K.-Y., & Zeger, S. L. (2002). *Analysis of longitudinal data* (2nd ed.). Oxford University Press.

Dormann, C., & Guthier, C. (2024). Longitudinal data collection. https://doi.org/10.4324/9781003362715-15

Dziak, J. J., Li, R., Tan, X., Shiffman, S., & Shiyko, M. P. (2015). Modeling intensive longitudinal data with mixtures of nonparametric trajectories and time-varying effects. *Psychological Methods*, 20 (4), 444–469. https://doi.org/10.1037/MET0000048

Epskamp, S., Hoekstra, R. H. A., Burger, J., & Waldorp, L. J. (2022). Longitudinal design choices: Relating data to analysis. https://doi.org/10.4324/9781003111238-12

Erez, L. (2020). Data structures for storing and manipulating longitudinal data and corresponding novel computer engines and methods of use thereof (U.S. Patent No. 10,740,345). U.S. Patent and Trademark Office.

von Eye, A., & Schuster, C. (1998). *Regression analysis* for *social sciences*. Academic Press.

Fahrmeir, L., & Kneib, T. (2001). Bayesian smoothing and regression for longitudinal, spatial and event history data. https://doi.org/10.1093/acprof:oso/9780199533022.001.0001

Fairbrother, M. (2014). Two multilevel modeling techniques for analyzing comparative longitudinal survey datasets. *Political Science Research and Methods*, 2 (1), 119–140. https://doi.org/10.1017/PSRM.2013.24

Ferrer, E., & McArdle, J. J. (2003). Alternative structural models for multivariate longitudinal data analysis. *Structural Equation Modeling*, 10 (4), 493–524. https://doi.org/10.1207/S15328007SEM1004_1

Firebaugh, G., Warner, C., & Massoglia, M. (2013). Fixed effects, random effects, and hybrid models for causal analysis. https://doi.org/10.1007/978-94-007-6094-3_7

Frazer, B. (2020, June). Creating and maintaining the person linkage spine. Methodological News (ABS Catalogue No. 1504.0). Australian Bureau of Statistics. https://www.abs.gov.au/methodologies/methodological-news-jun-2020#creating-and-maintaining-the-person-linkage-spine

Frees, E. W. (2004). *Longitudinal and Panel Data: Analysis and Applications in the Social Sciences*. https://doi.org/10.1017/CBO9780511790928

Fu, L., Wang, Y., & Wu, J. (2024). Recent advances in longitudinal data analysis. *Handbook of Statistics*. https://doi.org/10.1016/bs.host.2023.10.007

Funatogawa, I., & Funatogawa, T. (2018). Longitudinal data and linear mixed effects models. https://doi.org/10.1007/978-981-10-0077-5_1

Grimm, K. J., Davoudzadeh, P., & Ram, N. (2017). Iv. Developments in the analysis of longitudinal data. *Monographs of the Society for Research in Child Development*, 82 (2), 46–66. https://doi.org/10.1111/MONO.12298

Hall, D. B. (2007). Models for intensive longitudinal data. *Journal of the American Statistical Association*, 102 (480), 1473–1473. https://doi.org/10.1198/JASA.2007.S228

Hamaker, E. L., & Muthén, B. (2020). The fixed versus random effects debate and how it relates to centering in multilevel modeling. *Psychological Methods*, 25 (3), 365–379. https://doi.org/10.1037/MET0000239

Harring, J. R., & Hancock, G. R. (Eds.). (2012). *Advances in longitudinal methods* in the *social* and *behavioral sciences*. Information Age Publishing.

Ip, E. H., Pan, J., & Song, X. (2013). A bayesian analysis of generalized latent curve mixture models. *Statistics and its Interface*, 6 (1), 27–44. https://doi.org/10.4310/SII.2013.V6.N1.A4

Jensen, L. R. (2022). Using data integration to improve health and welfare insights. *International Journal of Environmental Research and Public Health*, 19 (2), Article 836. https://doi.org/10.3390/ijerph19020836

Kapetanios, G., Serlenga, L., & Shin, Y. (2023). Testing for correlation between the regressors and factor loadings in heterogeneous panels with interactive effects. *Empirical Economics*. https://doi.org/10.1007/s00181-023-02390-1

Kohli, N. (2017). Review of longitudinal structural equation modeling: A comprehensive introduction, by Jason T. Newsom. *Structural Equation Modeling*, 24 (1), 154–155. https://doi.org/10.1080/10705511.2016.1202772

Kunz, J. S., Staub, K. E., & Winkelmann, R. (2021). Predicting individual effects in fixed effects panel probit models. *Journal of the Royal Statistical Society Series A-Statistics in Society*. https://doi.org/10.1111/RSSA.12722

Ladbrook, M., & Hendrickson, L. (2020). Higher education completion rates for Aboriginal and Torres Strait Islander students using a multi-agency matched population cohort. *International Journal of Population Data Science*, 5 (5), Article 1491. https://doi.org/10.23889/ijpds.v5i5.1491

Liang, K.-Y., & Zeger, S. L. (1986). Longitudinal data analysis using generalized linear models. *Biometrika*, 73 (1), 13–22. https://doi.org/10.1093/biomet/73.1.13

Liu, H., & Chang, L. (2010). Multivariate longitudinal data analysis. https://doi.org/10.1016/B978-0-08-044894-7.00286-4

Liu, K., Saarela, O., Tomlinson, G., Feldman, B. M., & Pullenayegum, E. (2024). A bayesian latent class approach to causal inference with longitudinal data. *Statistical Methods in Medical Researchnull*. https://doi.org/10.1177/09622802241298704

Liu, S., Rovine, M. J., & Molenaar, P. C. M. (2012). Selecting a linear mixed model for longitudinal data: Repeated measures analysis of variance, covariance pattern model, and growth curve approaches. *Psychological Methods*, 17 (1), 15–30. https://doi.org/10.1037/A0026971

Liu, X. (2015). *Methods and Applications of Longitudinal Data Analysis*. Academic Press.

Liu, Z., & Guo, W. (2024). Longitudinal data analysis by hierarchical state space models. *Handbook of Statistics*, 109–153. https://doi.org/10.1016/bs.host.2023.10.006

McCormick, E. M., Byrne, M. L., Flournoy, J. C., Mills, K. L., & Pfeifer, J. H. (2023). The hitchhiker's guide to longitudinal models: A primer on model selection for repeated measures methods. *Developmental Cognitive Neuroscience*, 63 (null), 101281–101281. https://doi.org/10.1016/j.dcn.2023.101281

Moges, W., & Endalamaw, T. M. (2023). Application of longitudinal data the multilevel models approach on diabetes mellitus. *Journal of Health Science and Medical Research (JHSMR)*. https://doi.org/10.31584/jhsmr.2023936

Molenberghs, G., Molenberghs, G., Verbeke, G., & Verbeke, G. (2008). *Longitudinal Data Analysis*. https://doi.org/10.1201/9781420011579

Molenberghs, G., Molenberghs, G., Verbeke, G., & Verbeke, G. (2017). *Longitudinal Data Analysis*. https://doi.org/10.1002/9781118445112.STAT00291.PUB2

Mowle, J. (2020). Creating additional census income topics from linked administrative data. *International Journal of Population Data Science*, 5 (5). https://doi.org/10.23889/ijpds.v5i5.1482

Newsom, J. T., Jones, R. N., & Hofer, S. M. (2013). *Longitudinal Data Analysis: A Practical Guide for Researchers in Aging, Health, and Social Sciences*. https://doi.org/10.4324/9780203814208

Oberpriller, J. (2022). Fixed or random? On the reliability of mixed-effects models for a small number of levels in grouping variables. *Ecology and Evolution*. https://doi.org/10.1002/ece3.9062

Ong, P. S., Haußmann, M., Lönnroth, O., & Lähdesmäki, H. (2024). Latent mixed-effect models for high-dimensional longitudinal data. https://doi.org/10.48550/arxiv.2409.11008

Ilk, O. (2003). Multivariate longitudinal data analysis: Models for binary response and exploratory tools for binary and continuous response [Doctoral dissertation, Iowa State University]. Iowa State University Digital Repository. https://doi.org/10.31274/rtd-180813-11012

Peralta, Y., Kohli, N., Lock, E. F., & Davison, M. L. (2020). Bayesian modeling of associations in bivariate piecewise linear mixed-effects models. *Psychological Methods*. https://doi.org/10.1037/MET0000358

Raudenbush, S. W., & Chan, W. (1992). Growth curve analysis in accelerated longitudinal designs. *Journal of Research in Crime and Delinquency*, 29 (4), 387–411. https://doi.org/10.1177/0022427892029004001

Roder, D., Banham, D., George, J., Rushton, S., & O'Brien, T. (2023). *Demographic, health, and prognostic characteristics of Australians with liver cancer: A cohort study of linked data in New South Wales for informing cancer control*. BMC.

Santos, J. R. S. D., Azevedo, C. L. N., & Fox, J. (2021). Bayesian longitudinal item response modeling with multivariate asymmetric serial dependencies. *Journal of Statistical Computation and Simulation*, 1–36. https://doi.org/10.1080/00949655.2021.1965604

Saxby, K., Buchmueller, T., de New, S. C., & Petrie, D. (2023). The impact of employment on mental healthcare use among people with disability: Distinguishing between part- and full-time employment. *Scandinavian Journal of Work, Environment & Health*, 49 (8), 598–609. https://doi.org/10.5271/sjweh.4123

Song, X., Lu, Z., Hser, Y., & Lee, S. (2011). A bayesian approach for analyzing longitudinal structural equation models. *Structural Equation Modeling*, 18 (2), 183–194. https://doi.org/10.1080/10705511.2011.557331

Steele, F. (2008). Multilevel models for longitudinal data. *Journal of the Royal Statistical Society:* Series A *(Statistics* in Society*)*, 171 (1), 5–19. https://doi.org/10.1111/j.1467-985x.2007.00509.x

Tarling, R. (2008). *Statistical modelling for social researchers: Principles* and *practice*. Routledge.

Unwin, A. (2010). Exploratory data analysis. https://doi.org/10.1016/B978-0-08-044894-7.01327-0

Usami, S., Murayama, K., & Hamaker, E. L. (2019). A unified framework of longitudinal models to examine reciprocal relations. *Psychological Methods*, 24 (5), 637–657. https://doi.org/10.1037/MET0000210

Vermunt, J. K., & Georg, W. (2002). Longitudinal data analysis using log-linear path models with latent variables. *Metodología de las Ciencias del Comportamiento*, 4 (1), 37–53.

Wang, G., Wang, W., Mangal, B., Liao, Y., Schneider, L. S., Li, Y., Xiong, C., McDade, E., Kennedy, R. E., Bateman, R. J., & Cutter, G. (2024). Novel non-linear models for clinical trial analysis with longitudinal data: A tutorial using SAS for both frequentist and bayesian methods. *Statistics in Medicine*. https://doi.org/10.1002/sim.10089

Welsh, J., Joshy, G., Moran, L., Soga, K., Law, H.-D., Butler, D., Bishop, K., Gourley, M., Eynstone-Hinkins, J., Booth, H., Moon, L., Biddle, N., Blakely, A., Banks, E., & Korda, R. J. (2021). Education-related inequalities in cause-specific mortality: First estimates for Australia using individual-level linked census and mortality data. *International Journal of Epidemiology*, 50 (6), 1981–1996. https://doi.org/10.1093/ije/dyab080

Wijnberg, D. (2020). Defining populations within linked administrative data. *International Journal of Population Data Science*, 5 (5). https://doi.org/10.23889/ijpds.v5i5.1533

Zając, T., Perales, F., Tomaszewski, W., Xiang, N., & Zubrick, S. R. (2024). Student mental health and dropout from higher education: An analysis of Australian administrative data. Higher Education, 87 (2), 325–343. https://doi.org/10.1007/s10734-023-01009-9

Zapf, D., Dormann, C., & Frese, M. (1996). Longitudinal studies in organizational stress research: A review of the literature with reference to methodological issues. *Journal of Occupational Health Psychology*, 1 (2), 145–169. https://doi.org/10.1037//1076-8998.1.2.145

Zhang, P., Chen, X., & Sun, J. (2024). Regression analysis of longitudinal data with random change point. *Statistical Methods in Medical Research*. 9622802241232125. https://doi.org/10.1177/09622802241232125

Zhuowei, S. (2024). Regression analysis of asynchronous longitudinal data with informative dropout and dependent observation. *Communications in Statistics - Simulation and Computation*, 1–21. https://doi.org/10.1080/03610918.2024.2363952

Index

Note: **Bold** page numbers refer to tables and *italic* page numbers refer to figures.

V

W

For Product Safety Concerns and Information please contact our EU representative GPSR@taylorandfrancis.com
Taylor & Francis Verlag GmbH, Kaufingerstraße 24, 80331 München, Germany

www.ingramcontent.com/pod-product-compliance
Ingram Content Group UK Ltd.
Pitfield, Milton Keynes, MK11 3LW, UK
UKHW022123190926
475702UK00013B/1610
* 9 7 8 1 0 3 2 9 8 6 6 1 6 *